Certain Death in the Ice

Carl Lomas & Tracey Worth

www.istheworldround.co.uk

To Pauline

Certain Death in the Ice

Is the World Round?

Carl Lomas & Tracey Worth

The journey is the Adventure

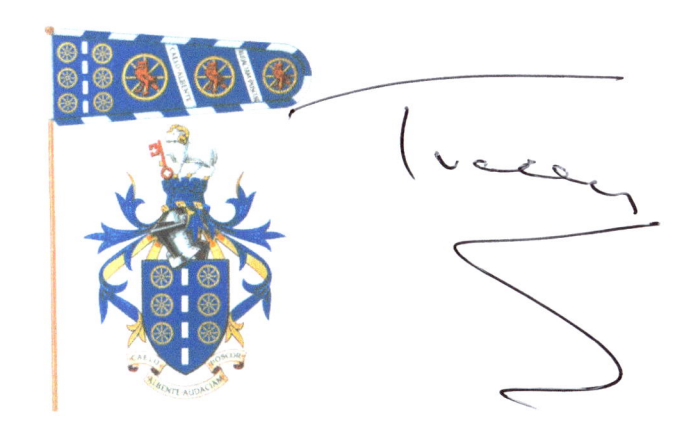

Tracey

Copyright© 2013 by Carl Lomas and Tracey Worth

All rights reserved by the author. No part of this publication may be reproduced, stored in a retrieval system or transmitted in any form or by any means electronic, mechanical, photocopying, recording or otherwise, without the prior written permission of the author.

ISBN: 978-1-905006-91-5

First published by The London Press, 2013.

Printed in the UK.

To the Royal Navy

*for saving our lives and the lives of our daughters
Caitland & Morgause.*

Other book titles in the series "Is the World Round"

Sunshine Days:
Blighty to Rio follows the family in sunshine sailing from the Mediterranean to the Canaries, Cape Verdis and Amazonia.

The Way Home:
Journeying from the ice to the Falklands, via Ascension Island, for the final homecoming on a big green plane.

Find out more information at the website: www.istheworldround.co.uk

Is the World Round?

The extraordinary story of a family's passion to engage the world.

They took an answer they believed in and set a question to make an adventure: *"We could never have believed the ending, but what joy to tell you the tale."*

To get the answer, it seemed a true price had to be paid; manmade canals would not do. Mystery, the end of the world and the history of sailing set a path on the ancient winds of trade and discovery. Not Panama or Suez but Cape Horn and the south to the roaring forties and beyond.

"From life in Middle England amongst the shires as far from the sea as you can imagine, all our lives we had been attracted to cities like bugs to the light — London, Paris, New York, finding the light was addictive but never fulfilled the desire, and the question built."

In "Certain Death in the Ice", you will join them in being confused by the ecology of green goo at Copacabana Beach and dazzled by the majesty of Sugar Loaf Mountain as they escape one more city. A roller-coaster storm propelling them into a run of mechanical misfortune amongst the light of more cities follows. They escape to see nature at one with God, but only to confront their greatest fears in an adventure for which the price to pay still leaves the question unanswered.

Is the world round?

Contents

Prologue xi

Prelude xiii

Chapter One
Running Beyond Rio 1

Chapter Two
Rio de la Plata – Uruguay 16

Chapter Three
Entering the Roaring Forties 32

Chapter Four
The Straits of Magellan to the Pacific 54

Chapter Five
Pacific to Puerto Montt, Chile 66

Chapter Six
Puerto Montt – Happy Easter 94

Chapter Seven
Robinson Crusoe Island – South Pacific 112

Chapter Eight
South for Fairway Lighthouse and a Right Turn for Magellan 124

Chapter Nine
Glacier Avenue, Puerto Williams & Cape Horn 136

Chapter Ten
Cape Horn bound for the Falklands 154

Chapter Eleven
Falklands Fun 162

Chapter Twelve
Falklands, bound South Georgia 178

Chapter Thirteen
South Georgia, bound North 198

Chapter Fourteen
South Georgia Repairs 208

Chapter Fifteen
South Georgia – a Second Exit, bound Cape Town 216

Chapter Sixteen
Realisation We are Sinking 230

Chapter Seventeen
An Asset Available, But It's Too late 238

Chapter Eighteen
Hollinsclough Sinks 250

Epilogue 257

Acknowledgements 259

Annex 1 – A Tour of the Yacht 261

Prologue

Carl and Tracey began life on the water with a small motor cruiser on the Thames, followed by years of French summer cruising from Dieppe to Cherbourg, with trips via Shotley Point to the Dutch lakes making for autumn adventures. Their daughters, Caitland and Morgause, grew up on the water from the tiniest boat-show life jackets to surfboards in Biarritz and warm-water snorkelling on equatorial reefs.

Although power boaters at heart, ocean travel outstretched the distance of diesel range, so they looked for an Oyster sail yacht for a blue-water solution, and Rupert Knox Johnston sold them their 55 cutter-rigged sloop.

Transferring their powerboat electronic navigation systems to their sailing yacht, they drew on their thousands of miles of motoring experience to rely mostly on sail power this time to travel on the old Cape wind routes, without using the Panama and Suez canals.

Travelling the blue-water oceans, Caitland and Morgause commuted from boarding school in England to enjoy ever more remote adventures. Well placed before their GCSE years, they joined the yacht in Buenos Aires, planning to stay for a whole year for the Pacific leg to Australia, but it would not go as expected.

Thirty-six hours after impacting growler iceberg debris below the 50th parallel in the Southern Ocean ice convergence zone of the South Atlantic, the yacht would sink, and Caitland and Morgause would help their parents fight for the lives.

Prelude

Sailing in the sunshine

Sail yacht *Hollinsclough* left England in the early years of the new millennia. Leaving Ipswich for a South-Coast run across the Irish Sea to Douglas, Isle of Man for the 100th celebration of the motorbike TT — 130mph laps and an island party. At the ready were solar panels, wind turbine and blue-water cruising gear, all bolted on.

We headed south to France, down to Brest, then made a left turn for summer on the shores of Biscay. We took in the Glénan Isles, the stones of Carnac, La Rochelle, the sand dunes of Archachon and Biarritz for the surf. Basque carnivals and bullfighting were the destination of 270 west for the Spanish Main. Santander, Bilbao and then La Coruna took us south to Portugal for Lisbon and Cascais.

The Algarve, Cadiz and a salute on our trumpet to Nelson at Trafalgar took us ever closer to a new continent. Tangier, North Africa meant dates and camels before a turn north across the strait to Gibraltar. Restocking with English food supplies of rice pudding and baked beans at Morrison's,  we spent Christmas in this English place followed by two New Year's in one with a walk across the Spanish border for the hour change.

This was the start of our Mediterranean taster. Benalmádena, Torremolinos and then to Malaga to meet the Dakar Rally that was not to be. Then it was onward to the Canàries. Days of carnival followed in Tenerife with a mountaintop snow photo on Mount Teide. Gran Canaria was next for volcanoes and Fuerteventura, a barren desert, sent us all west for the whistling mountains of Gomera.

Going farther south, the Cape Verde Islands brought giant tuna and sail-fin marlin monsters in the water. Setting ever onwards, a spring crossing of the Equator saw Portuguese Men O' War floating by in purple splendour. Fifty degree temperatures of daytime and wild light in the dark evening ocean came from phosphorous plankton.

Heading for a South American summer against the big currents, we arrived at the island of Fernando de Noronha, swimming with Nemo and the turtles. We crossed the last of the Atlantic to the north Brazilian coast of mangrove rivers leading to Cabedelo. The mud rivers there were full of manatee, and monkeys and crocodiles lived all around. Following the coast south to Recife, we travelled on to Salvador for armed guards in the marina and maintenance for our yacht.

We arrived Abrolhos Bank off the coast of South America for the mating season of the humpback whales. We used an Abrolhos Islands anchorage for remarkable diving in the reefs.

The blue mafia, a name given to the "Girl Guides" around the world for their blue uniform, are friends as close as family with our young daughters the world over, and in Rio, we say hello to the Brazilian Chief Girl Guide Lucia Tavares Ramos and the International Commissioner, Maria Olinda Luz, and ask them the question… "is the world round?"

"Is the world round?"

Is the world round?

Wherever we have travelled, it is the people who have made the distance so special. Parrots and penguins, camels and monkeys, marlin and whale, glacier ice and volcanic fire — every day is an adventure.

Chapter One
Running Beyond Rio

Christ Redeemer statue Rio

"*The five am sunrise brought morning light and beautifully filtered turquoise water…*"

"*Travel was always about people to us.*"

"*Tired tormented and tortured by the majestic strength of nature, we gave in…*"

Brazilian city life and shops with goods at 110 volts were no match for life on an English 220-volt yacht.

Our daughters, Caitland and Morgause, had been with us for the English summer. Travelling by air, they had brought the gift of an electric toaster to replace an old, broken one. Now with steaming toast for breakfast, it was time for the girls to return to school while we focused on escaping Rio. We were bound for Uruguay in an October southern summer sailing run; we would collect the girls for Christmas via British Airways in Buenos Aires. From there, they would take a year out and, family together, we would sail the Pacific.

The bay of a thousand islands beckoned to tear our Velcro from the light of city life in Rio. James Bond movie locations were all about our quest to answer if the world was round, and today it was the turn of metal teeth smiling in the South American sunshine as Jaws waved goodbye from the cable car of Sugar Loaf Mountain.

Together and alone, we were on our way again. We slipped out of Gloria Marina, Rio de Janeiro with the five a.m. sunrise for the first Saturday morning of October. The green gloop of Guanabara Bay passed by as we turned about Copacabana Beach in the shadow of the mountaintop statue of Christ. Surf, sharks and city sewage was the reality for one of the most famous tourist beaches in the world. Copacabana was a contrast of paradise and manmade waste as unbelievable as breakfast without toast. It was as unimaginable as not returning home safely to England aboard *Hollinsclough*. Rio was a diverse, energetic paradise of South America, but it was no blue-water paradise in the ocean.

As the Atlantic waves, soft tide and gentle winds drove us south off the coastline, the rock gateway to Rio closed before the chain of enormous grey mountains faded into the mist. We were back in the ocean. Sea debris was gone as the morning air warmed the memories of a magical city in our minds.

We cleared the small, scattered islands of Tijucas and put 270 on the

Cable cars in the South American sky

autopilot steering dial for a westerly run. Marina goo was exchanged for fresh seawater. Sailing talk always ends with toilets, so we hand-pumped the porcelain bowls with fresh, ocean water, and we were once more at sea.

Our twelve UTC position posted as 23.03.30S 43.18.40W. A few miles on, we passed the Bay of Banderitas, the South American name for the Guides and Brownies. We had very fond memories of pack meetings in the city; we had been made so welcome in Rio, met the country's chief guide and been introduced to Rio Guide packs city wide that made us more welcome than anywhere else in the world.

Our plan of attack after leaving Rio was a soft day sailing to hold up in the Bay of Islands 60 miles west. We would use the anchor chain to settle back to the movement of the sea and complete a few small maintenance jobs such as cleaning the propeller and scrubbing the hull clean of its city life. The pilot books promised a cruising jewel of the south coast, but our target was just a pit stop before heading south to cover miles and put distance on the travel clock.

The day's wind was so soft we were unable to set the mainsail still; the softness of the wind left it flapping and lost in the air. Polly Perkins, the diesel engine, helped as we motor-sailed down the coast for a reintroduction to life on the water.

As the mist broke, the mighty coastal mountains tried to hide in the shimmer of the hot sunshine. The horizon line was edged yellow with sandy beaches that would surely be more credible than the green goo of Copacabana.

Turning back towards the coast from the day's passage, we passed the Laje de Marambaia light on a solitary rock island three miles out. Castelhanos Point then focused our attention for a run into the island Ilha Grande. The island was almost uninhabited, and it so reminded us of our wonderful Bay of Salvador adventures during our last leg from the north. A steep wooded jungle of vivid green rose up a few thousand feet out of the sea and touched the sky where mist clung to the cloud. The sky vista was broken by black vultures circling above red falcons.

We edged into the first bay of Ilha Grande around the Palmas rock to set our anchor off the beach of Mangues. A late afternoon delight in the silence of the forest, Smith frogs called out between the harmonies of insects that were interrupted by the shrieks of monkeys. Leftovers in the fridge magically phoenixed into a very English beef stir-fry before bedtime. Falling asleep amongst this green paradise, we were confused by the contrast so unimaginably close to the chaotic city of Rio.

The clear blue water of the Bay of Islands gave us the opportunity to snorkel and dive below the boat — clean of sewage, grease and debris — to polish up the propeller and break barnacles from the bow thruster. With 220 volts on the generator, we began the day's proceedings with toast for breakfast in the rainforest world. Jobs

CHAPTER ONE: Running Beyond Rio

Paradise devoid of mankind

were a plenty, with all energy in the warm water before the heat beat the hands of the clock to turn life into a lazy afternoon contemplating a job well done.

The energy of city life behind us, we took a breath a short dinghy ride to land on the beach and explore this paradise devoid of mankind.

We followed a muddy red-sand track in the hillside, crossing the small island to view the Atlantic. This was uninterrupted ocean from Cape Horn in the south. A crescendo of energy crashed into the yellow line of sand, which squealed in resistance. The landside of the sand was edged with windswept palms where bamboo shoots six inches in diameter anchored in their stand to the wind, a wind we wanted to turn so it could carry us on our way.

A small green airstrip brought some rare access to claim this beach as one of Brazil's finest surfing haunts. We stood on the sand of Praia Lopes Mendes, which swept at least a mile in its defence of the waves. This was no giant Hawaiian roller coaster, but the length of the white run was impressive enough to encourage any roller-coaster addict into one more ride.

Waiting for the wind turn, we found jobs to fit in with the wash down of the morning rain. Clouds sought land in their dance with the low pressures of the changing weather. Nature had made this island so much the greener with warm rain; today, the warm rain washed our wetsuits clean of salt water.

A second afternoon trip found us on a northerly route march above the bay. Climbing a steep rocky path trodden by surf tourists, we twisted ever upward into the steamy canopy of cloud mist. Giant blue butterflies danced in the green, the cry of frogs rang about our ears, and the shriek of vultures with the rasping of bugs made life cool enough for the exertion.

People but no roads

The view from the top was a vista of South American elegance. Turquoise green water in curvy bays of yellow sand, green canopy hillsides all framed with damp, white mist — steaming water airborne back to the clouds from whence it came.

We were shocked to find the island did have mankind. We could see the town of Abraão in the distance, and noticed that a total absence of roads isolated it from all but those in a boat. From our mountain view, we could see a busy water taxi service running about the local islands. Boats from small craft to tourist schooners shuffled about their business. Moving tourists to remote beach bed and breakfast posada houses, the logistics of food and life was all at sea. The next island had a small church and many houses painted mallard blue — the paint clearly only came in one colour.

Within three days, our Velcro was torn from the addiction of city life, but unable to tear morning jobs from our city schedule, we began a review of the standing rig of the mast. While ropes do all the flexible work of the sail, strands of steel wire ping the mast tight to the yacht and put the big strengths into the deck. Each steel cord, made of many stainless steel strands, has a base adjuster to take out slack as mileage builds and the temperatures change. With only half a turn on the adjusting bar, the port rear mid mast stay broke two of ten strands of its stainless cord and signposted a small turn in our adventure.

So much for just a pit stop.

A haven of charter yachts based at Angra lay in the shelter of an island 20 miles to the north. Piratas Mall Marina awaited, and George and the team were happy to see a world cruise yacht divert to arrive in their tourist marina. *Hollinsclough* was in good hands. George found an engineer called Jamie, who was a specialist rigger and expert steel man. The offending steel was down in a jiffy and away for re-fabrication. Fourteen millimetres pulling about seven tons of force out of the windward side of the mast.

Back to mankind, we chilled for the following day in the lazy two-street town and did some shopping. We found a short piece of plumbing pipe that would make an excellent sheath for the anchor rope, new blades for the hacksaw in a local merchants, and the finest local lure in the fishing shop. If every local is to be believed about their fish lure, why do they always get away? We had a Bob's burger lunch, as Brazilian McDonald's was a shore-based treat. There were many water taxis and fishing boats on the town quay, and a number 73 bus was ready to commute locals home on the more conventional roads of this larger island.

Jamie returned the following day with the shiny new steel smartly coiled. It was soon bolted to the mast and ready for action, but Jamie wasn't happy. "Your mast is twisted. If you are heading south, we must make it straighter," he informed us.

It looked pretty good to us, but with Cape Horn on the horizon, Jamie had his mind set on a perfect job, and we were not about to argue. Together, we spent a whole day tweaking, tightening and loosening the side stays, back stays and boom weight. Going 60 feet into the sky with the aerials, it is truly astonishing how much flex and twist the main mast has when its stays are loosened up for adjustment. A hundred hours later in a single day, pinging tight on the hydraulic rear stay adjuster to bring the cutter tight, and the job with 47 cups of tea was a "good un". Did we mention wire-brushing all the adjuster threads, greasing up the bolts, cleaning down the threads, new split pins in every hole and a healthy spray of silicon for a shiny finish?

The whole rig, mast cutter, jib and boom finished close to where it started, but it was clearly so straight the double bubble in the level became one, and everyone on the pontoon had something to say about the cool lines of an English sloop cutter in the evening sunlight. With this job done, we would sleep easy in any gale we may face to the south.

A modern marina inside the paradise islands

The steel had been a good omen, and the jobs earned our right of passage, as nature turned the end-of-week Friday wind for us. We tore the Velcro of Rio clear for good and cast the lines from Piratas Mall.

We left the island of Angra running near dead south on the wind. Clearing the Ilha Grande to our port, we were on the east side of Meros Point and beyond the big ship anchorages before lunch.

The spinnaker pole was rigged starboard, and a healthy wind catapulted us into the sea with a third mainsheet set and a full cutter. The seas built through the afternoon, and the darkness hid 20-foot walls of wave to our port stern. We ran the same sail set during the night for nine knots, bagging a ten point five on the ground speed dial to show just how well that rigging had been adjusted.

A few ships ran very close to us out of Sao Paulo, one even setting the AIS (Automatic Identification System) avoidance alarms off. Then a mysterious beast with a single twinkling red light stood before us. The light configuration of just one red confused us. Ten miles closer, and we could see it was an oil rig, white lights topped off with a single giant flashing red.

Somewhere in the early hours, 25th parallel south-ish and charging along at ten knots in the darkness, one of those sea walls said "I am a rogue" and hit back. *Hollinsclough* took her first pooping. A million gallons of rogue wash caught on the downside of a double set of big waves poured in over the stern. Confusion and surprise obliterated reality with a dark disorientation of wet, loud and forceful water.

However, *Hollinsclough* ran her course without deviation. Tracey and I took the bump of a heartbeat while physical consciousness felt the first sensation of standing in a bath of cold water. The second sensation was the force and speed of it all. The third sensation was pain. I was no longer sat in the cockpit but pinned against the canopy steels, harness tight around my chest magnifying the urge to gasp at the cold dark air.

Nature had spat back. After city life and lazy days in the islands, the rogue wave over the stern in the darkness was a wake-up call to say hello and make sure we knew we were bound for the end of the world in the south.

Hollinsclough was an English Oyster, and they build them well. The water from the rogue wave went down the twin floor drains in a moment, but there was a casualty — we had long grown accustomed to leaving the inner kitchen window open. The window high in the galley kitchen wall sat in the base of the cockpit. It never got wet down there? Yes, chef, it did in a rogue stern pooping, and it would be a lesson learnt for the future. What a mess, and what a mopping up job.

CHAPTER ONE: Running Beyond Rio

We rose at around five a.m. and gave us a view of the water and what had grown into a full gale. White spray tore above the teeth of waves that engulfed the view as we tore down what felt like a roller-coaster run. The compass danced in its three-dimensional ball, signposting the only realty of which way was up. Tracey shouted for a log reading, and I was grinning when I replied, "it's about 180 south plus or minus, say 30." The dial danced a tango to the wave roll of our lost horizontals at such a pace you could have called almost any angle you wanted.

The forecast softened by lunchtime, and the sunshine was building. Our 24-hour ground target was about 170 miles — shaken but not stirred. Damp but not drowned — "not going to die today" — it was time for a cup of tea as we sat together in the cockpit, proud of the mileage in what really was our first day back in the ocean after too long in the city of Rio.

Our Saturday morning position for 12 UTC October 11th was 25.25.50S 45.28.40W, running south to surf some of those walls — big sea country was building. Roast beef and two veg for lunch? Don't you believe it. Our sea diet when the waves turned big had long adjusted to the skill of boiling water and getting it into the half-torn top of a Pot Noodle cup. The cup shook itself in an automatic mixing ceremony during silver service delivery for feeding in the cockpit. The cup went from burning hot to stone cold against the lips and was held with blue fingers.

Pot Noodles four times a day was our food on the run while sailing down the wind. Any leftover hot water made it to the tea for dunking dry biscuits for afters.

The winds continued to soften, drawing down the speeds and returning life to the centre lane of the big blue one-way street south. It was a light evening, courtesy of a full moon. The soft cloud diffused the shimmer and revealed one of the most magical phenomena of the ocean. The 25th parallel south of the equator marked one of our last sightings of this magic as the water around us shone with phosphorous plankton. It was only a little and long gone from the equatorial elegance of warm waters to the north, but it was enough to generate momentum to our dreams of warmer times in the land of King Neptune — the centre of the Earth.

After a number of years of sailing, we had adopted six-hour watch patterns and left behind more conventional short sleep three-hour blocks. Between the two of us, we greatly preferred six-hour sessions — midnight to six, six to midday. For us it worked well, I took the midnight to six a.m. stint and went to bed with a Pot Noodle breakfast that usually followed the sunrise.

That evening found two more oil rigs with their eerie single red lights. The depth gauge marked shallow 100-metre water in the mighty depths of the South Atlantic for the end of my night shift. The five a.m. sunrise brought morning light and beautifully filtered turquoise water below the white crests of the dark blue ocean.

Our weather window out of Brazil for Uruguay was changing against us. The Ugrib download forecast a big low sweeping into our path 300 miles to the south. We were three days out from the Ille de Grande bay, cautiously aware of the just-back-to-the-water feeling after a month ashore. It was no time to stand a 30-knot wind on the bow for a pasting and a big tack.

We took the splendid wind that had given us 180-mile-a-day averages for the option of a soft move west back to the Brazilian coast. A 40-strong pod of small spotted dolphins agreed with our decision as they came to play in the arc of our turning wake. With giant leaps and loud splats, our swirling tummies were forgotten as we laughed and smiled at this flotilla of fins. Florianopolis lay 100 miles down a beam reach sail — a lot of mainsheet out for another splendid run on the wind.

A night-time arrival amongst the local fishing boats always focuses the senses for collision, but add floating oyster beds and we were down to tiptoe entry speeds for the anchorage. A crescent-shaped bay amongst the islands of Laje Moleques and Ilha de Coral made perfect protection.

With a few circles, still a little confused by the fish penlights, we aired on caution and chose to put the anchor down in the north of the horseshoe crescent called Enseada de Pinheira. The Christmas tree lights of fishing boats were outshone by the bed and breakfast porch illuminations on the beach, and the light show became a twinkling army watching over us in the darkness.

Back home, school half term was beginning, and the girls began their half-term holiday without us in the safe hands of former Hollinsclough head teacher Mrs Wherry. The satellite phone was not just for the turning weather but for reports of the first few days of school holidays. For all of the big holidays, the girls would fly out long haul to be with us. Rio had worked well, and Buenos Aires was ahead, but on this sailing leg and with a short half term, they would stay in England to sample life at home from boarding school at Repton.

Our morning half-term sunlight in the South Atlantic anchorage rose to display a white sand edge around our new crescent home. The oyster beds bobbed about in the gentle water, well protected from the open ocean. Mighty mountains climbed up and disappeared layer by layer into the distant mist. Between us and the north beach lay floating marker lights of the oyster beds bobbing about in the turquoise water that had been so confusing in the darkness. Locals worked them from small rafts, and tourists from the Pousadas bed and breakfast chalets walked the beautiful beach of palm tree postcard paradise. The Ugrib weather files proposed a three or four-day wait as that low moved across to tear up the open Atlantic.

You can take city folk out of Rio but not the city out of the folk. Drawn to the light, we were not hard pushed to make a decision to spend the wind turn inland. We set sail, motoring the Polly Perkins diesel engine 20 miles up the Canal Sul de Santa Catarina for a few days in the city of Florianopolis. The mountains marked our track on either side, wind funnelling down at 20 knots for a frothy drive up shallow channels narrowed by ever more oyster farms. Obstacles of desire made this a daylight-only run to unfold the enormous bridges connecting the mainland and Santa Catarina Island, where Florianopolis sprawled out on both sides of the water. We watched the depth gauge fall on approach, two metres closing to zero for the last mile. A bottom of grey gloopy mud sucked and spat silt from the propeller wash. Tide in our favour, there was still nothing left on the depth gauge to get into the marina.

We anchored in the centre of the river, well short of the motorway bridge in the light and sound of the city. We could have been off Tower Bridge London but for the absence of water traffic. It was almost a week since our feet set foot on land. From tranquil beach anchorage to the sprawling urban metropolis, we sat in *Hollinsclough's* open cockpit and absorbed the evening lights of the city. Sound, light and sensations were so close but so distant by our detachment from a pontoon. The water tanks were full, the electricity from the solar panels and wind turbine had the batteries charged, the cooker was on diesel and we were blue-water cruising down town with not a need in the world.

The following day was a splendid relaxation in the cockpit. Salt, dried from the ocean spray, glistened in the sunlight as strongly as the city lights had in the night.

Our log entry was "Florianopolis Bridge South at anchor 12 local 13 Oct. Position 27.36.46S 48.33.00W All well."

How is it that no sooner is one settled than the world changes? The predicted low ripped up the South Atlantic, turned nasty, expanded and spat its intent inland. Safely anchored in the shallow river, we watched the gauges go up to 28 knots against the tide. The current accelerated its speed in the narrows towards the motorway bridge, and we had a gale blowing around our little city anchorage.

Not surprisingly, the anchor began to drag. "Put some more chain out," Tracey said, smiling. *Hollinsclough* would always side to the wind when she dragged, and today was no different. Her decks were uncannily stable, but the motorway bridge was still closing on our position. I looked into the chain locker to find I already had every link out. That was 50 metres of chain in three metres of water. It was not going to hold!

Heading to Cape Horn and the Magellan anchorages, we were well organised for a drag. With few words said, we went into action. Polly Perkins, the diesel engine was fired up — hydraulics turned on — the motor helped turn the bow back into the wind, and we started winding. Hornblower would have been proud; it was textbook stuff, and we were underway.

Meanwhile, Florianopolis Yacht Club members enjoying lunch decided that the launch of their rescue boat was the most sensible action: *"The Angletere boat is awful close to the bridge,"* they must have decided. Bless their cotton socks. The rescue boat was a small, shallow-draft ship tug. I banged my head on the chain locker lid as I looked up to see it roar across our beam, hooters howling with a seaman in smart whites throwing a tow rope at me with inspiring precision as the captain steered into our lee. I was quick to fasten the rope tight against the forward cleats, and the tug took hold, standing firm into the wind. He was so keen to complete his rescue that there was hardly time to reel in the anchor chain.

"Steady as we go." This was down-town excitement. The VHF tug radio was on digital transmission for yacht in distress. "Thunderbirds are go!" The whole city launched. We were listening VHF radio emergency channel sixteen, but it was all in Portuguese.

The harbour control launched, and their rescue RIB (Rigid Inflatable Boat) joined in a race with the lifeguard RIB to our beam. "Steady on boys!" It was all exciting stuff; we were in safe hands, but we needed a minute with this anchor chain.

The small ship tug had the local knowledge to swing out of the river centre to an indistinguishable dredged channel not on the chart that delivered us to the security of the tiny pontoons of Florianopolis Yacht Club. A world touring yacht was clearly a rare visitor in the tiny marina.

Newspaper reporter Gabriel from *Noticias de Día* was waiting for our arrival on the fuel quay.

The club's tug let its rope go, and the world and three dogs awaited our lines to tie us up in a spider's web to the most posh pontoon in town. As the building wind whistled around us, the most English voice in the world boomed out, "Take a few days." It was the club commodore.

Our addiction to city lights found us once more Velcroed to the city. We had no complaints after a week at sea; the Velcro was secured with side-to mooring for the easiest life ashore. We had mains water aplenty, meaning endless showers and the washing machine on, plus broadband speed wi-fi to remind the girls we were still alive in the South Atlantic hardship of city life.

Florianopolis Celebrities

We spent a week in the City of Florianopolis, southern Brazil, to match our week in the ocean. The wide motorway suspension bridge illuminated our evenings with coloured lights that washed into the cityscape of vibrant energy all around us. There were pedestrian high streets with endless shops — all individual, no chain stores here — two covered markets and a vegetable hall. There was fresh bread and cake at streets stalls, cobbled walkways under tall palm trees with 22 degree sunshine, warm and not too hot for a suntan, which camouflaged us as locals.

Newspaper interviews in the cockpit

    Camouflaged we may have been, but bridge celebrities we were. Gabriel, the local journalist, plunged us full across the back page of the city daily.
    Blimey, it was more a story about travelling the world than a touch with that suspension bridge. Smiles afloat, pontoon fame and near rock-star status in the yacht club… it was a whole lot of fun to be there! Travel was always about people to us, the warmth and friendship grew with the distance we travelled, and Florianopolis was no exception.

Dinner invites and deck discussions as yacht club members practiced their very best English warmed our motivation for the quest for our answer: "Is the world round?" We were being fattened up with food and friendship for a trip south. Each evening, midnight blue crane birds lined the pontoon as we walked home to our cabin in the river. Masters of the pontoon pylons, they formed a guard down the floating boards of our Brazilian shoreline home. *Hollinsclough* was in such good condition it was hard to find jobs, but the engine room got a splash of bright blue paint on the Polly Perkins engine hoses. We fitted some new air fans to the kitchen and some smart low-energy LED diode spot lamps to the mast courtesy of the last B&Q delivery from the girls. We sluiced the bilges with fresh water, scrubbed the decks and sang "me 'earties" as onlookers grinned at the crazy English abroad.

Outside in the ocean, the South Atlantic had settled. The monster low tearing it up had passed, and it was time for a last sailing chums dinner invite. This meant a car trip north of the island to the edge of the freshwater lagoon, which was framed by beautiful white sand dunes. A Brazilian meat feast barbecue followed with Angela and Fabio, and their daughter Anna, Carlos and his daughter, the neighbours, and Daniel, who gave us a midnight lift back to the marina.

Even with such warmth of friendship, our Velcro had to be torn away. We returned down the shallow waters of the canal for a night's anchor back in the bay of Enseada de Pinheira amongst the oyster farm that had so confused us in the darkness on our first arrival. Courtesy of the electric toaster, we enjoyed a hearty breakfast and, without Sunday morning newspapers, made our way south.

We posted our twelve-lunch position at 27.57.30S 48.28.30W 26th October and settled in for the run to Uruguay. We set half the jib, all the cutter and a third of the mainsail for a beam run in good wind and soft swell.

Predictability is not a strong point of South Atlantic forecasting. Our four-day run turned into six as the wind unexpectedly turned all about and another new low pressure roared down upon us. Top of the gauge was 36 knots, but worse still was that it turned and came in at 40 degrees to the bow, making it hard work to hold any reality of a course. The Brazilian coast south of Rio had shown us a whole run of very unexpected and unpredictable wind turns. The turns were so enormous that you could see them in the sky as they formed above to spit back and send you away.

A true monster formed above us, and the blades of the aerogenerator screamed as the wrath of nature circled into shape for an onslaught.

CHAPTER ONE: Running Beyond Rio

The wrath of nature circled above.

Heave to or run? In a bid to push on, we chose to tack as tight as we could physically bear to the wind. We had an intent to make ground, valuing it at all cost to comfort. We had a splinter of main sheet sail and half a storm sheet on the cutter. Rollers clearing the bow came over the canopies to repeatedly drench us with ocean and drain all momentum. Open ocean mountains crashing around us. The horizon was gone, the monster engulfed us, and the sky was lost to a whirlpool of grey that came at us from every corner of our vision.

For over six hours, we had 40-knot winds 30 degrees off the bow, and we bounced from one sea wall to the next, only making a single nautical mile of ground to the hour. The mast howled and the hull replied with deep pain, the sandwich of deck rubber to metal mast twisting against the forces of nature to turn the stomach in anguish of forthcoming disaster. The mind racing with visions of structural failure, the only comfort came from the work done with new steels. Every bar had been adjusted to perfection on the mast fittings, but would it survive today?

There was unspoken warmth as we hugged tight together, and we felt an inner feeling of comfort to know the girls were safe home in England. It was clear that should any piece of the yacht fail, we would be in perilous danger. It was the first time in all our ocean adventures we had faced the reality that "We could die today."

The harness lines were tight on our bodies, and our feet were locked against the cockpit walls, which had become the floor. Feet pinned fast, this was our tightest position, under the shelter of the canopy, which was doing a masterful performance of survival as the onslaught of the storm tried to batter it down.

Our motivation to hold course had been drowned in the waves. We had sailed half the world, but the unpredictability of gales turning in this piece of the South Atlantic defeated us. It turns, turns some more, and as you look into it breaking on all sides, it is shocking to reach that moment when you can hardly predict the direction of the anger that nature has spat at you. Wave sets break into rogues that can't be there, and a froth of anger roars up to explode with snarly white teeth. Tired, tormented and tortured by the majestic strength of nature, we gave in.

In our minds, there was no failure in dying here, but we swung the autopilot angle away from our destination. It was heartbreaking to be over 100 degrees off course before placing the wind to the comfort of the stern quarter. The seas still lost in their direction on the port side, we chose open ocean to landfall. With just a hanky of storm sheet keeping steerage and stability on the cutter, *Hollinsclough* roared off to take rocket-speed action. Catapulting down the circling waves, the ride was wild but alarmingly manageable.

We were heading backwards to our course on a one-way street; the no entry signs flashed by, and the ground was lost. We were off the wind, but that relentless hammering of the waves had gone. The mast still groaned, but the hull stopped creaking.

A sleepless evening later, we found enough energy for the preparation of a hearty Pot Noodle. The ocean hadn't yet found enough kindness to provide dessert. Saving enough hot water for steamy tea and dunking biscuits was beyond our ability to walk on the kitchen walls. We tried to balance the mugs to our lips without washing the head-linings of the galley in this roller-coaster world of tumbling, tormented waves hunting a direction, but it was impossible.

We fought in the teeth of a torn-up storm to turn back for two days. Wind turning and tacks tightening, the autopilot dial almost surprised us as it began to show enough south to drive the belief that we wouldn't end up back in Rio.

The Pot Noodle spillage lining the cockpit was cleared twice over by two enormous poopings over the stern of the cockpit — our kitchen windows closed today. Down below was awash with kitchen spillage but dry of saltwater. We would find humour in draining the water from our rubber boots in the bathtub of the cockpit as it exited those double drains, and that humour signposted that the battle was turning in our favour.

As we naturalised to this incredibly uncomfortable environment, a last night gave two shorter tacks that brought us to a bearing over south, and we were heading for Uruguay again.

The morning sunrise drenched us with motivation for life, but we wouldn't quickly forget our first sight of mortality when nature chose to spit back. Somehow, the ocean was more beautiful for experiencing its anger.

So much ground was lost that the leg doubled in days at sea. Tired, wet and tormented by the wind direction, the soft current at last yielded, and nature allowed us into Uruguayan waters.

Having fought the storm, we were given the greatest reward of a wicked evening wiz down the wind into the shelter of the Rio de la Plata. We had earned this water with all our energy. It had been our first fight for our lives, and it was probably the hardest ocean battle we fought in all the time ahead.

Punta del Este beckoned, a Uruguayan Whitbread stopover for world yacht racing teams. That would do nicely. Shaken but not stirred, today we were alive, and the city lights beckoned.

Chapter Two
Rio de la Plata – Uruguay

Magellan looking over the city & pointing the way downwind Buenos Aires.

"No matter where you travel, the world is made smaller by the people you meet."

"An English Red ensign flag flew on the courtesy pole to greet us."

"A statue of Magellan pointed to the sea…"

From the Rio de la Plata turning into the shelter of the bay of Punta del Este, we arrived in darkness to take the safe option of the mooring buoys. Chango and his boys dashed out of the marina in an RIB to help tie not one but three mooring lines to a white open-water mooring buoy in the harbour entrance to the lee of the small island of Gorriti. Safely in the bay of Bajo Los Banquitos, we could see the lights of the city and slept well after our stormy ordeal down the southern Brazilian coast. With 8,000 miles on the dial, sea dogs we were. Shaken and stirred, with growing log miles, the rollers were just getting bigger all the time down here, but then summer was coming.

By morning, Chango's team returned and piloted us into the marina of Punta del Este. They then tied us up bow too with stern ropes to running buoys. This was no spider's web; they shackled, not tied, the ends, and even lock-wired the shackles. Blimey, it blows that hard here? I asked myself.

The port is a stopover for the Whitbread around the World Yacht Race, so it was no stranger to foreign travellers. Rolex and Musto was the norm and was enormously welcoming to real sea dogs. At Punta del Este Yacht Club, we were welcomed by Commodore Horacio Pastori, signed in as visitors from the Royal Chanel Islands Yacht Club on our royal blue ensign flag and made to feel very much at home. A club of colonial splendour, it was lost in time and pampered members to rest in elegance in a land that time forgot. We were an unimaginable distance from the storm we had just weathered.

By Sunday, some of the boat was drying out from the South American storms, but more importantly, it was time for a birthday lunch — three courses and club claret to wash down the giant portions of roast beef. But for the daylight, we could have been in London east of Grosvenor Square for a candlelight super in the Saville Club in Mayfair. It was a long way to Claridge's, and there was no low-calorie option. Our addiction to the lights of the city was not to be disappointed here.

Top world sailor Jim Kilroy invited us to dinner with his wife Nelly, and there was much talk of his *Kiolia* yachts. *Kiolia Three* held the Sydney to Hobart race record for a staggering 21 years. Discussions began about the Irish Sea — a parking zone for the Fastnet Race? It was during yacht club lunches in Punta with Jim that we would be taught the art of heavy-weather sailing from firsthand experience. Jim would talk of 100-knot wiffs, and however bad it was, you needed to get power into the back of the mast. They were rules we were taught to live, but sat there in Punta, we could never imagine such adventure happening to us.

After Florianopolis, we had got a grip of this travel business with celebrity status. The newspapers loved our story as the miles unfolded, but here we stepped

The look of a Saint Bernard

further into stardust with a visit from Canal Ten TV. That's the BBC by South American standards. A snazzy Panasonic movie camera with Leica lens on a Manfroto tripod wooed our eyes. Happy to let us speak English, they said the subtitles would go on afterwards. We must have been a little foreign in our localness. Tony had signwritten everything on the boat. We called him back in Blighty to send out new world tour signs for the boom; "*World Tour*" no "*El mundo viaje*" — Spanish of course. "We are coming back — promise."

Our best friends in Punta del Este must be the sea lions. Monsters by any standards, the giant beasts sun themselves on the quaysides, rolling about the concrete looking more lion than sea as their hair dries in ruffles and turns African. The beasts take an easy lunch with the fishermen, who throw leftovers in a double-bubble deal to get the tourists buying their wares. Mammals, descendants of land creatures believed to be large dogs, they returned to the sea a million years ago. So, up close and dangerous we get to find a startling likeness to the face of a St Bernard. Almost local, we wait for them to turn, then run past at full speed with our shopping from the Uruguayan Disco food store. A hearty roar quickens the heart as our blood rushes to the legs for the quick dash by. No time to hang about.

Guy Fawkes and November the fifth was absent in this faraway place, but November 14th is Uruguay Army Day, and it was good reason for *Hollinsclough* to raise the coloured courtesy pennants of the countries she had visited. What a sight! A Marine officer and a naval commander were very impressed as they toured the harbour by patrol boat. The Air force and the navy compete for resources, and the coastguard is also a military service here called the *Prefectario*. The word is that Blighty donated almost all their Hong Kong helicopters to the *Prefectario* before a little more of the Empire slipped away.

A splendid day finished with a dark delivery. The Marines doing night sorties in their stealthy black RIBs scared us and the giant seals to death as they popped out from under the quayside in the dead of dark.

CHAPTER TWO: Rio De La Plata – Uruguay

Punta del Este had a magical uniqueness. Sat on a long spit of land, it was both rough and protected all at the same time. The well-protected marina bathed in the Rio de la Plata inside of the spit was only a short walk to the Atlantic Ocean side for a fun-in-the-sun, surf paradise of consistent rollers.

Needing exercise, we signed up with the local surf instructor, an athletic chap called Gherman . Richard at the surf shop took a look at our short board and smiled. "Save that for the youngsters," he said. Three lessons on long nine-foot boards had us standing and thumbs up for a photo. Then another storm rolled in to rip up the shallow water of the estuary and, like sailing, put our boarding on hold.

We retreated for maintenance — the jobs of storm run oceans never end. While in Punta, we took stock to find a supplier of anchor chain, and with the help of a split link, we turned our 50 metres of ten-millimetre anchor chain into 100 metres. Surely that would have been enough for Florianopolis bridge.

Londres, Lisboa and Paris, it's never the same name beyond your back door. The Rio de la Plata is better known back in Blighty as the River Plate. This is where the Graf Spee was scuttled by the Germans. The Ajax and Achilles closed in to block the pocket battleship's escape into the South Atlantic like the storm that was behind us. Not history buffs but in love with films that follow our travels, we ordered up the movie. We never thought for a moment that it would be in black and white.

The Uruguayan capital Montevideo has half the country's population, but the port is shallow, with limited room prioritised for commercial shipping and no marina. Swapping sail for wheels, we took the hour's bus ride down the long straight toll road.

Montevideo was a meeting with the blue mafia. Girl Guide celebrity news had followed us from Brazil. The International Guide Commissioner, Aiejan-

Montevideo Parliament with Uruguay Guide Aiejandra Varela

dra Varela, of Varela Zarranz wine fame, was waiting for us. She did the most fantastic job of showing us the city. Guide frocks on and toggles shiny, we had a VIP tour of the Parliament building and the Legislative Palace — a chamber of 30 senators down the corridor from 99 deputies in their party colours. The place was guarded by the Battalion Florida, a smart household guard protecting the 1825 constitution in a giant hall of marble, stained glass and gold leaf. A public library was at the centre of the building with every book published in the country. It was reminiscent of the old British Library, but here the building was square and so wooden you could have been at the Bodleian but for the lack of manuscript chains.

The lights of the city dazzled. Modern shops were down the high street, with Fendi, Valentino and Chanel boutiques, and modern art sculptures centred on a giant hand in the sand for the tourist photos.

We went all cultural to see the Theatre of Solis, an opera house with one level of seats before a wall of floor to roof boxes. Then it was fresh air and sunshine to visit the Rose Garden in the enormous municipal park. It is a wonderland capital city hardly the size of a market town in the shires of England but adorned with ceremonial buildings that are mostly modern of Italian architecture dating from independence in 1825.

We ate meat for a celebration dinner. Everything for everyone on a giant platter char grilled in barbecue ovens before your eyes — a feast of splendid proportions.

Back in Punta, the ocean was alive with athletes.

The snipe sailboat racers had arrived for a week of fun in the Western Hemisphere sailing championships. We rallied behind Canada and Bermuda as old colonies of British Empire and stood to attention on the VIP line up for the opening

Racers from the western hemisphere await battle.

ceremony of flags. Two navy captains and their commanders stood with the Captain of the Port and Commodore of the Yacht Club in best blue blazers. English abroad, hats on and white trousers for a weeklong party sailing extravaganza that was nothing short of Cowes Week.

This was real sailing.

We motored our small RIB alongside the wind of competition, and then spent a day in a big RIB with the judges and two more on the commodore's motor yacht.

The greatest technical problem for November was the absence of rich fruit English Christmas cake. None of the like was in the shops here, and don't even think about mince pies. We found brown sugar and fruit everywhere. Cook and Bligh might have shipped some aboard before they set off for the South Seas, but our Fortnum supplies had long run out. It was time to measure the biscuit tin and rally sailing master Roy Dunning at mission control in Blighty for a marzipan recipe.

Christmas was closing, marina space in Punta was at a premium for the summer sunshine holiday season, and the Commodore invited us to New Year's Eve dinner. But first we had to visit Argentina and think about collecting the girls from their next flight out.

Brown like dark-sugar marzipan, the Rio de la Plata estuary stood before us. A wide shallow estuary which narrowed to an inland ship canal dredged deep for shipping traffic to Buenos Aires in a line up like cars in a London traffic jam.

Hollinsclough returned to her travels for an easy inland run down the Rio de la Plata estuary. We left Punta in soft winds for a 170-mile leg to the Uruguayan city of Colonia. The sea lions eating their lunch wondered who had just gone by. With whiskers the size of our spinnaker pole, they shook their heads, thought of Christmas and went on fishing.

Twelve ships sat in a line anchored off Montevideo awaiting a place in the port. The soft wind forecast wiped up into a 30, as forecasts do, and the VHF radio rattled out 40-knot gusts warnings. We had many memories of the Dutch coast runs to Ijmuiden in choppy water made far worse by its shallow nature.

The brown water turned white and frothy, and we strapped down for a roller coaster run on the surf with a full cutter and even some jib. Shallow water funnelled into estuaries can tear into the teeth of a sailing trip better than any deep-water storm, and the brown goo hiding its mysteries with the depth gauge showing shallows added to the anxiety of a dark night out.

Many ships bound for Buenos Aires run the narrow channel for Argentina on the south side of the estuary. Bulk ore carriers, a car transporter, many tankers and a few thousand-foot container monsters were with us that evening. Our AIS avoidance computer reported everything in a long dark watch of dancing lights on this river ocean. Flat lands without mountains gave little respite from the howling winds, but our speed was good.

The coastguard called us to check all was well 30 miles off the Argentinean coast, 8,400 miles out of Blighty. They couldn't stop wishing us good luck. A military twin-rotor Chinook helicopter took a fly by, and seven navy vessels passed our bows. It was endless water, but all focused in shallows with nowhere to turn. More than 20 ships were anchored outside the Buenos Aires container port. The ships were like monoliths off Easter Island, all sat into the wind, watching and waiting for their turn.

With the wind building but no open-ocean exposure, we exchanged Pot Noodles for baked potatoes. We mixed them with butter and tuna and yum-yummed our way to the port as the white frothy water liquidised around us.

Buenos Aires lay 20 miles to our port side, but as our paperwork was still Uruguayan, Colonia lay to starboard. We called Colonia port control on VHF16 for our entry, identifying ourselves as English sail yacht *Hollinsclough* and proudly formalise with our international call sign VQRT8.

"Wait," she replied. "I call back shortly in Engliesh!"

Bless their cotton socks. We rounded the island of San Gabriel, watching port buoys 100 feet from the shore in a tide race to make Alderney proud. We used manual steering and bow breakers to speed the heart for a tight entry turn into a soft bay.

Our position was posted at 34.28.00S 57.51.00W Colonia del Sacramento, Uruguay.

We had a day to be tourists and get our travel papers stamped, and with the Argentinean city across the way, there was still a local to meet in Uruguay.

Colonia is an old town of Roman-like grids that could be from the dark ages but was actually founded in 1680 with cobbled streets and tourist shops full of flags and matte tea cups. There was a bull ring, the fights long gone, a supermarket with tinned corned beef and an Antel phone shop for ET — fone "ome! We also found an

Argentinean embassy and some remains of the old city wall. We enjoyed a local Chivita lunch of a giant steak sandwich and chips, and spent a lazy afternoon in the town square of palm trees under the statue of Artigas. The evening passed at the aquarium museum — catfish and piranha.

Colonia del Sacramento is town of lazy days and old cars with brown water sea that left you reminiscent of Windsor. Henley wasn't up the river, but the lights of Buenos Aires beckoned just across the estuary.

There were many like it, but this one had no MOT.

No matter where you travel, the world is made smaller by the people you meet. We found a yacht under a Namibian flag with a Scottish owner, Ian, on board. Ian introduced us to his German wife, Doris. Friends of the grandson of Baden Powell, it was the talk of the blue mafia and the Girl Guides until the wee small hours. They had visited Matlock back in Blighty ten miles from our home. Ian had some great advice on deep-sea disasters. A big flood problem, turn the engine sea cock intake to "off." Cut the pipe and rev the engine to use as the best pump on the boat. It was a wicked idea not to be forgotten.

Buenos Aires – December

Papers to change countries stamped and passports cleared, it was a soft run in the shallow brown water estuary of the Rio de la Plata to move *Hollinsclough* from the Uruguayan tourist town of Colonia del Sacramento to the metropolis of Buenos Aires. We made six knots with an acre of sail in nine knots of wind for a sunshine trip between the ship channels. Dredgers did their relentless duty in the ten-metre lanes, keeping the bottom clear for the giant cargo ships. Fast ferries buzzed about the bay, and pilot boats ferried experts between buoy pick-ups to navigate ships down those tiny lanes. The AIS collision computer triggered our location.

At 11:30, December 5th, the Argentinean coastguard welcomed us. *"Sailing vessel Hollinsclough, welcome to Argentina."*

Golly, that's 8,400 miles out of Blighty, I thought.

Entering from the Canal Norte into the ship basin, we were surrounded by glass-fronted tower blocks. A yellow barrage moved back on its hydraulic ram to let us into the marina of Yacht Club Argentina. Chums back in Uruguay had organised a berth for us, and an English red ensign flag flew on the courtesy pole to greet us.

To clarify how hot it was, lizards near the size of dogs walked the grassy banks of the marina. They could easily have passed for a T-Rex on a diet in the haze of the hot sun. Terrapins bobbed in the shallow brown water of the city's nature sanctuary.

The yacht club was downtown London Mayfair, wooden panels and Victorian paintings of ocean, ships and sea battles, marble floors and fresco ceilings, and dinner was formal. It was a sort of English gentleman's club abroad with a foreign language and a seaside location.

Small world and all, the English flagged wooden ketch *Nordwind* sat on the fuel quay. Skipper Michele and his wife Penny welcomed us. We hadn't seen him since Rio, but then that was hardly 1,000 miles ago — distant sail travel is a very small community. They were exchanging a washing machine, and what stories we both had to tell of 110-volt conversions for the South American appliances plugged into English boat generators of 220 volts.

Buenos Aires marked Christmas school holidays and a return to family sailing. The girls safely arrived from Heathrow by British Airways on the longest single leg long-haul in the book. Whilst we had fought the storms, they got a tour of the cockpit, and the pilot offered them a Mars bar each, which sounded better than Pot Noodles in a gale to us. It was strange for the girls to leave English snow and arrive in Southern Hemisphere summer.

Chocks away for regular air miles unaccompanied

Buenos Aires was child heaven with a McDonald's on every block, Burger King too, but no KFC. The proud capital city flew, soft blue Argentinean flags fly everywhere, and this may be the best dressed city in the world. Ladies wear dresses, not skirts, and men sport jackets even at the most casual of times. The cabs are black with yellow roofs, and petrol stations marked YPF serve Renaults, Fiats and Peugeots

CHAPTER TWO: Rio De La Plata – Uruguay

that hardly fill the five-lane roads. The city boasts buildings of glass, a large smart city zoo, a four-line underground train and a fine art gallery the size of the National back in Blighty — they even have a Turner!

The Natural History Museum had dinosaurs, but it was nothing on our live marina lizard show in the hot afternoon sunshine. A war memorial in the plaza of St Martin was so reminiscent of Paris we could have been in Europe, but this war involved the Malvinas and the map did not read "Falklands."

Malvinas or Falklands?

The Retiro district has an English clock tower and sports shops busting over with bouncy training shoes. They call them "tackys", but we think they mean "trackies" — the American slang conversation of the Spanish back to English might gotten a little lost. Small malls of designer stores hide below tower blocks, and boutiques of European fashion sit comfortably beside antique stores, cheap grocery shops and expensive bookstores.

Posh street markets reminiscent of Covent Garden sit next door in Recoleta district and sell leather handbags and jewellery between panpipe music and the smell of barbecued food.

A number 152 bus played tango music as it headed west bound for the district of Boca. A transporter bridge marked the gateway to girls in stockings, their pants shown in the twirls.

Football was in dark blue here. This was not the Gunners and the Emirates Stadium, this was Boca Juniors. This was land of Maradona. The football stadium was in the shape of a chocolate box coloured golden yellow and royal blue.

It takes two to tango.

25

Morgause in amongst it

On the terraces

The penalty of this faraway place was the heat. *Hollinsclough* donned her mosquito nets in the 30-degree sunshine.

A blue mafia day in best togs followed to meet the international Girl Guide director, and then a Saturday VIP reception with 170 guides and brownies on full parade for a ceremonial salute to world chums. We were lost in the centre of a big, blue ocean of friendship.

A statue of Magellan pointed to the sea, toward the cool air of the coast. We enjoyed a last Argentinean lunch of ham and cheese as more folk danced the tango in San Telmo. Buenos Aires is a city of polo, tango and tourism, but with the heat, we could bear no more.

A blue ocean of friendship

CHAPTER TWO: Rio De La Plata – Uruguay

Out of Buenos Aires

For the last days before Christmas, we sailed back to Punta del Este, making heavy weather of the River de la Plata estuary. The forecast was good, but the tide and current fought against us. The distance placed us across two tide turns to watch the escape of the river battle the South Atlantic. It was a rougher run back than we had taken inbound, and the shallow water all ripped up

River Plate traffic jams

with nowhere to go crashed about the decks to wash down the girls' breakfast. We turned a little more north to follow the fast ferry route to the Montevideo channel. Everyone was hunting for depth, but having ships so close had become normal to us as we chilled in the sunshine.

We saluted the Graf Spee buoys and then sailed beyond Buicio for lunch in the lee of the Ilha de Flores for an evening arrival on the familiar mooring buoys of Punta.

We spent Christmas in Punta del Este, Uruguay, the town filling up with tourists and the shops opening. It was so busy that the council turned the traffic lights on.

We arrived as amateur body boarders and became short-board surfers in the hands of Gherman at the surf school on that magical beach on the far side of the marina spit. Daily lessons had us all standing in the soft water. Caitland was the first to ride the seven-foot-six board we had carried on our whole adventure. Surf chums down the beach enjoyed the moment as we did.

Christmas was a hot sunny day, cruise ship *Infinity* was parked up for a paparazzi photo, and the yacht club was open on the 25th for afternoon teas.

Caitland stands on a seven-foot-six board.

27

We had a date with the steep sand dunes. Morgause was surfing the sand courtesy of a sand board from Santa.

The yacht *Nordwind* arrived for a December 31st reception with best frocks and then a New Year dinner at the yacht club, including lobster starters, sweet and sour lamb with pork for the main course and white chocolate pudding.

Midnight here was four hours later than in Blighty, and we delighted in text messages bouncing about the hours of time zones. A firework display of epic proportions launched across the bay to mark New Year in Punta, and we set sailing resolutions for the year ahead.

Christmas Day is not the same in sunshine.

"After heading south from Uruguay, the next big landing will be Easter Island."

After New Year, there was to be no delay taking the winds to move us down the South American coast.

A full restock of food involved shopping delivered by van with more than 100 Pot Noodles, spaghetti and tinned peas stuffed to bursting in the lockers. We took our last mail post from the yacht club and registered our forward route for the next boat to bring any late Christmas cards for delivery to us at Easter Island. We checked out of immigration, said goodbye to the sea lions of Uruguay and sailed down a soft wind with a surfboard adorned with the autographs and best wishes of memory.

Lunch out of Punta was off the island of Lobos. A nature reserve seal colony, the water chums screamed out with joy as they swam on their backs, fins up like raised sails, floating in the waves for fun. We made two days and two evenings of very comfortable wind under a wild, dark sky of Milky Way stars. The sea was rich in dolphins, who marked the sunrise with an ocean dance about our bows. Scrabble, card games and maths homework

Santa's sleigh swapped for a sand board!

changed the daily sailing focus from a two-watch team to a family crew. We hooked a sea mackerel and stored him in the rope locker approaching port at Mar del Plata — 38th parallel south.

Luis and his Yacht Club Argentina port team took our ropes on a rickety wooden pontoon set, pulled us in and tied us up. There was good wi-fi, 50-hertz electricity and water pipes in the base of the pontoon timber boards. Life in harbour never got better than this. The sea mackerel made a fine feast with our feet on the shore.

Our surfboard was an autograph book of memories.

Mar del Plata claimed to be the foremost beach resort in Argentina, and it did not disappoint. Children in the sea schools sailed their single-mast Optimist dinghies, a diving platform floated in the clear blue water, and ice-cream kiosks provided refreshment between beach games and body boarding. A fine swing bridge made for adventure treks fit for Indiana Jones. Great territory for the dinghy and pilot jobs to be done helping race yachts onto mooring buoys. Morgause honed her dinghy skills, dodging sea lions in the water to ferry folk from boat to shore. She was in training for Patagonian rope mooring work.

We had a blue mafia day to meet the 136th community group of Girl Guides. The marina yacht club workers were so overwhelmed with the link to their community that we couldn't escape without lunch.

Fed and watered and back in the hands of the blue mafia, we met Area Commissioner Stella Colantonia, who awarded Tracey the Argentinean friendship badge. We sipped Matte tea and ate local cakes with our new chums while swapping badges, games and songs.

It's fish for tea then.

Mar del Plata marina team

We were the English guests in Mar del Plata, and a Union Jack adorned the courtesy pole of the club lawn in our honour, as it had in Buenos Aires. A stream of visitors came to meet the family travelling the ocean. Diego Perkins invited us to his live-aboard yacht, and he explained that the yacht club had 2,000 members but only 100 boat owners, and only two people lived on the water. He organised a slot on the tennis courts for us, and we yielded to the red clay speed.

Visitors from Buenos Aires Jose and his wife Gracelia on yacht *Cosa Nostra* drove us past the Mar del Plata Naval Base with its grey submarines, minesweepers and frigates.

McDonald's stood tall above the breakwater in Mar del Plata along with two surf schools, a Tango club for little ones like a French Le Mickey club. The golf club had its own sand, tables, café and wind awnings. There was a polo club too, the Ocean Club, and of course Yacht Club Argentina.

We joined Diego and his boys at the surf school for more Punta-like waves — long, low and soft. it was heaven for easy-run learners. Progressing with gusto, we had all left the soft boards for hard fibreglass and were

Surf and skateboard balance never came from dry land.

heading sub-seven while Morgause fully mastered the balance roller to become the surf shop face of English balance.

We had hardly seen a world travelling boat on the South American continent and then two English sloops arrived in a day. A wooden classic *Merrymaid* that so reminded us of *Nordwind* and a modern boat called *Festina Lente* with Nick Pochin and his chums aboard. Nick was on a mileage mission and heading for Cape Horn to turn for a run north to Vancouver. He had already circled the world in a staggering 20 months. It was a world of sailing apart from our adventure tangled up with friends, lifestyles and city stays.

The knives are sharp for a meat feast.

Caitland and Morgause rallied chums, sharpened knifes and sorted a barbecue for the adults.

Nick was focused for a direct run to the Falklands, but storms building around Cape Horn in the south delayed the window of exit for the seven to eight days needed for the run. It was possible to port hop the coast in shorter legs, but the passage books and even local knowledge were very thin on the route.

Hollinsclough was a little north of the roaring forties, and ready for Orca. At 8,400 miles out of Blighty, would the next leg be to the Falklands? Or would we coast-hop Darwin's route via Puerto Madryn and the Orca killer whales?

Friends and sailors together

Chapter Three
Entering the Roaring Forties

A family of sailors. Puerto Madryn

"*This was the land of friends and memories.*"

"*Welcome, Welcome. The friendship of faraway places...*"

"*If it was the end of the world, we needed to get a good night's sleep.*"

It was mid-January and we were well behind schedule to run Cape Horn. Our life was focused on weather forecasts, watching both the coast run and the Falklands sea leg. Would we bag the Falklands prize or hug the coast for a port-hopping trip? Either way, Cape Horn and the end of the world lay ahead.

The boat was well, only needing a good clean. A light propeller shaft rattle meant changing the propeller end anode with the help of a local diver. The anode is sacrificial, rotting away with corrosion as the weak link. Held by three bolts, it starts to rattle as the corrosion grows. The first time it did this in the Mediterranean, it scared us to death, and we thought the gearbox was about to blow up.

January in the Southern Hemisphere was summer and fun in the sun. With each day of unsettled winds forecast, the Velcro was tightening to the port of Mar del Plata, but we watched eagerly for a weather window.

The big question ruling our lives was whether to coast-hop via the Orca territory of Puerto Madryn and Puerto Deseado or take a seven to eight-day ocean run to the Falklands. The prize of a Falklands visit was a big one, but that left a hard push into the wind to get back to the Strait of Magellan. Coast or Falklands was the water discussion of every moment in the marina.

As we familiarised ourselves with the weather patterns, the days surfed by. We took lazy walks to the beach resort of Mar del Plata. The whole fam-

Their boards are a whole lot shorter than ours!

Caitland chills with the girls.

ily was now on hard surfboards, and Morgause was chilling with the surf instructors as she faced the bigger waves with ever smaller boards.

Yacht Club Argentina had been a friendly place, their pennant hung on our visitors' stay, as did a Union Jack on the club pole ashore. In a language match of foreign places, we had placed Spanish *"El mundo viaje"* on the boom where it was marked in English, *"World Tour"*. Multilingual — it was a truly multinational affair!

With the weather reports stable on the seven-day forecasts, it was all go for the Falklands. Nick Pochin on *Festina Lente* was first out for the ocean run.

Marina captain Luis and his staff made sure our send off was just as good as the stay. With the marina staff in their launch and friends on the pontoon wobbling about as they waved and cheered with voices heard to say "See you in England", we followed *Festina Lente* out of the marina.

Both yachts took a short but heavy sea swell to motor out of harbour before we had building wind behind us to sail for the Falklands.

Ten miles out, however, our sails were not working well. Lifejacket and harness were on for a visit to the bow to check out the problem. The cutter rig had sheared the base spigot on the stay. It could be hand furled — rolled in by hand one turn at a time — but we would never be able to fix the hydraulic driving spigot at sea. The decision was mixed. We had waited for what promised to be a great weather window, but safety first swung the answers: *"Back to port."*

Luis watching us with surprise, he called us on the radio for news. Our berth and friends awaited our return. With a team of hands, we had the cutter down and the tools for the repair waiting. A local machine shop lathed up a new spigot, and by the next morning, we were shipshape again.

The weather window was great for Nick, but for us, another low now five days out closed our target of the Falklands.

It is so easy to look back and the cutter would have probably set without a move for the whole run, but our attitude had always favoured repair than make-do. It's a tough call to let a weather window slip, but that was that, and we were back in Mar del Plata watching weather patterns yet again.

CHAPTER THREE: Entering The Roaring Forties

We had almost twelve-hourly downloads of weather forecast Ugribs to judge and re-judge. With each day, we became experts at watching the change of the forecasts, but seven-day windows eluded us. The season of wind was not going to wait for us.

Leaving Mar del Plata for the second time, we were not heading for the Falklands. With sadness in our hearts at having to take the coastal route, we were heading for Puerto Madryn, and that warmed us with its reputation for colonies of Orca whales.

Our second send off was as warm as the first. The seas were calmer, the winds were lighter, and a slippery black harbour seal waved us around the port wall. We turned 180 degrees south to meet friends on the Argentinean yacht *Cosa Nostra*, who were fishing on the Pescadores Bank. A frantic exchange of greetings at sea and promises to see each other in another place followed. With one last wave, we swung down the wind with 240 degrees on the autopilot dial, the sails filled with air, and we were in the footsteps of Magellan, Darwin and Drake on the roaring forties coast run of South America in search of the end of the world.

We were hugging the very rugged and sandy shore of the Argentinean coastline. The *Prefectura* coastguard had already called us for our MMSI (Maritime Mobile Service Identity) number, and we were now in their hands with twelve-hour email postings of positions. Back home Falmouth Coastguard was watching our EPIRB, and our land-based sailing chum Roy was monitoring positions and weather for the English yacht deep down south.

There is very little in the pilot guidebooks for this coast, details of the forward ports, diesel and stores are thin even by local standards. We would find the first cartography map mismatch to the GPS co-ordinates, we started recalculating the curvature of the Earth in relation to our position.

We were mostly on the sail, but surprisingly the shaft rattle was still there when we ran the engine. Had the diver not tightened that anode properly? The soft, well-set winds gave us the time to enjoy family Scrabble but with keen minds on catching the double letter or triple word score, it was still fit for an ocean battle to the last. The fishing line was out but there were no takers today, so mashed potato and meatballs followed by stewed apple kept us going till bedtime.

A wild red sunset, the like you only ever find at sea, and a biblical night of Milky Way stars followed. The manmade world was still with us in the soft glow of lights from the city of Miramar to the north as we dozed in the warm cockpit of *Hollinsclough* at 39 south.

The Roaring Forties

At six p.m. UTC on Thursday 22nd January 2009, Morgause blasted the ship's hooter at full volume for a deafening good enough for the artillery. "*Hold fast!*" The occasion marked was crossing the 40th parallel south, the numbers were all cool, 60 west and close on 9,000 miles on the log out of Blighty. In standard terms, that's right out there on the edge of the world and well beyond Greenwich zero.

Forties on the dial, we approached the Golfo San Matias with a 500-foot cargo ship, the *Auckland Star*, closing on our stern. It was the first of a smart run of tankers in and out of San Antonio Oeste. The sea was very kind, and we had almost all the mainsail out with the boom well swung and the jib poled for soft downwind sailing where the stars were not the only excitement.

Deep into the short hours of the second night's watch, the wave breaks made the wrong sound, and shrieks and whistles followed. *Hollinsclough* had startled a large pod of dolphin in the dark night. We turned on the underwater lights, beaming into the depths of the South Atlantic Ocean; playtime was wild with school out as torpedo-like dolphins shot across the waves. It was a dazzling display of dance the like of which we had never seen before.

Daybreak at 41 south, and temperatures were falling quickly. Fifty miles of the Valdes Peninsular found more firsts. We sighted five Peale's dolphins — smallish chaps with short black noses that were almost beak like and brilliant white undersides. How they reminded us of collie dogs of the Peak District, darting and dashing about in decreasing circles of fun, not for sheep but a hearty fish lunch. We made do with a fine Pot Noodle, steamy and easy for life in the roaring forties of the South Atlantic.

The lowering of the huge, perfectly formed forties south circle of the sun passed before our eyes as yet another sea-bound sunset approached. The night time closes in quickly, and it is time for coats and navigation lights.

It was the first time since the Canary Islands that we had seen a passenger ship come up on the AIS avoidance computer. The *Silver Cloud* was just that. She too was heading for Puerto Madryn — we had a sailing chum. It wasn't long before we saw a bright white glow in the distance as we watched the screen and saw the ship closing our stern. At ten miles, we could make out the three ladies on the top deck having a very good party. *Silver Cloud* had more lights on her than the *Queen Mary*, so we put our sail lights on to join the party. A radio call followed. "*This is English yacht Hollinsclough, Silver Cloud. Over.*"

"*Are you having a party onboard your yacht?*" called the watchman of *Silver Cloud*. He bid us a fast and safe sail and looked forward to seeing us in port.

CHAPTER THREE: Entering The Roaring Forties

By sunrise, Morro Nuevo point was in sight. We could see what looked like the white cliffs of Dover, but it was Punta Ninfas. *Hollinsclough* was coated with a grey yellow dusty mixture of grit solidly attached by the warm salt spray to give us a Gulf War veteran look. The currents were kind to us for a low-water arrival that let us into the mighty bay without resistance. Sadly, there was no sign of the Orca pods we had expected. This was a place void of all signs of mankind; no telegraph poles, no houses and not even the trail of a jet plane in the sky.

We eased north of Punta Loma in soft wind for a shock. Out of nowhere, in all this shelter, the wind came about hard from the west making 40 knots plus close into our bow. After ten miles of motoring, Polly Perkins, the metal sail, was fighting nature for a run to the anchorage.

We were motor-sailing in a perfect match between engine and wind with great urge to get anchored before this weather turned into a monster. Tacking tight zig-zags with the cutter sail and the engine full on, we made an astounding five knots over ground. Dancing from port to starboard on the sail, that shaft rattle from the engine side of the effort became a heartbreaking grinding sound. This was clearly no anode problem.

The last miles were a slow, hard-fought battle. Our energy level was weak, but then the sight of our friend the *Silver Cloud* on the commercial dock beckoned us to the goal of the anchorage. That old adage that you have to earn these legs was true today. In that last run up the bay, we had taken inch by inch ground to arrive hand steered, wind at 45-plus and 30 degrees to the bow for a last push into the anchorage.

The girls stowed in bed below, we screamed "tack" time and time again to grab every inch. Winch pulled in, then we hand-tightened to get the very best set on the sails with intent not to lose any ground.

The wind was straight off the beach like a no-entry sign for a one-way street. The girls, aware of the yacht slowing, popped up to see the excitement. The sea was tearing up surf from the beach with a haze so white the town was unclear.

"Can we anchor in this?" Caitland asked.

With a big smile I said, "When the anchor bites, it will be torn back so hard it should stretch the chain."

Polly Perkins went into neutral, reverse kicked in like a rocket motor, and the anchor went down like a brick. It dug in so tight we thought the chain would pull the winch out.

There is nothing like a solid chain to count sheep for an easy night's sleep after an afternoon's work on the winch handles of the cutter sail.

At anchor in Puerto Madryn

In the morning sunrise the beach gale had softened, and we headed for Puerto Madryn, Patagonia, Argentina — get the postcards written. *Hollinsclough* anchored in the roaring forties, land of Patagonia, land of Right Whales, Orcas, dolphin and albatross.

Madryn was founded by Welsh settlers in the late 1700s. It was a desert-like land of rolling sandstone hillsides. There was no marina, just giant pylons for the cruise ships.

Back to that shaft damage. Checking the boat, we found metal debris by the gearbox head of the propeller shaft. The shaft thrust bearing had collapsed; failure of the thrust bearing had damaged the CV joint linking it into the gearbox. *Ouch!* With credit to the design, it had a force surface to hold it when the bearing failed so had kept running metal to metal to get us into port.

That said, this age-old Welsh settlement was no marina, so we anchored in open water. Beach landings were a very wet business, thrown in on the surf from the seats of our motor RIB dinghy in white water fit for the log flume of Alton Towers. A French motor cruiser out of La Rochelle, the *Bourniquel*, lay here at anchor. Laurent, a Dakar rally man, stood with me above the damage. He looked down and offered advice on the engineering in best French: "*Ahh shiit!*" The famous French sailor was headed south with his wife, Caroline, but had time to help us understand the design and get the bearings apart.

You take marinas for granted until there isn't one. Repairs at anchor are no sideshow, and lest we forget, we had past 40 south. The CV joint and shaft apart, we had no facility to turn the prop on the motor. We were in the hands of nature at the sole safety of our anchor.

The weather turns that hadn't given us our window to the Falklands came in to say hello again. The beach wind built up to 50 knots, so the next answer was to lay our second anchor. There was help from the yacht club for this. A big RIB bounced in the waves to drop our second chain to three runs of rope, giving another 100 metres of line. It was perilous work with both lines on snatch ropes and on buddy weights pulling down the lines. Everything that could be done was done to hold us tight. The *Armada Prefectura* were on notice in case we dragged, but the lines held strong, and *Hollinsclough* held fast. Do these moments make for a good night's sleep? Don't you believe it.

In the following days, we were befriended by everyone. Caitland and Morgause made the Optimist sailing boat school of the yacht club their home. On one rough night, they stayed ashore with school head Valeri, just like life with Mrs Wherry back in school holidays too short to come out to the yacht.

Fifty-plus gusts in the anchorage were common place here, and the locals took them in their stride. We found it easier to sleep in the cockpit strapped to the harness lines ready for action. To be in bed below was beyond us. Who would ever go to the Southern Ocean?

A second night of the big winds was too much to bear. For the very first time on the trip, the weather placed us all ashore for a family evening off duty in the Hotel Muelle Viejo for a landlocked breakfast of hot bread and warm milk far from the 220-volt toaster on the English generator.

Away from the boat, new chums even found the girls some new clothes. The *CNAS Club Nautica Atlantic Sud* did everything possible to help with the bearings. The club commodore taxied us about the town's engineers. Boat engineer Lois moved us onto the workshops of a precision machinist called Paulas. Dive boat captain Daniel Remenar spoke good English for our translations, and so did Frank, the director of the sailing school.

It was a world without part numbers where engineers carried vernier gauges and machined metal with a skill passed down from their fathers. We placed the girls at the lathe to watch an art we had forgotten back in Blighty.

Teresa in the club café fed us heartily whenever we were on land and had free toffees waiting for the girls whenever we were ashore.

The daily newspaper "*Diario de Madryn*" did a two-page report on our travels, and by the following day, food and gifts were arriving at the yacht club for the stricken English yacht under repair.

A descendant from some 19th century Scottish settles called Albert arrived with cakes homemade by his wife Tuto. We were soon at his home for afternoon tea. Golly, the whole town was on hand to

Turned by hand without part numbers

help. There is no doubt in our minds that Argentina is the friendliest country we have ever visited. Can you imagine if a foreign campervan broke down in Derbyshire? Would all and sundry drop jobs and run to its aid? The friendliness and helpfulness of the local people here put us to shame.

The small town of Puerto Madryn was far from the main world of Argentina with many gravel roads, but to our benefit, engineers from another generation still lived here. This was not a land that had galloped into an age of convenience; there was no *"call for a part and change it."* Here everything was repaired. The thrust-bearing plate was re-worked by hand on a lathe. A large new wheel bearing from a Peugeot van was used as a replacement for the old thrust rollers. With dexterity and skill long lost in Europe, the engineers stripped the CV joint. They found a Land Rover differential bearing to be a close match. They only had one, but it was reworked to a diff bearing from a Ford pick-up. They handmade a centre spline to match both bearings and re-worked the whole joint to create a new one in two days. They strutted about, vernier gauges in hand with a head full of experience for what would work in diversity. The cost was around 200 dollars. Meanwhile, it took Roy back in Blighty most of a fortnight to source a new unit for 1,000 UK pounds, but that was almost 10,000 miles away in a different time zone. There were no part numbers here.

Girls celebrating the engineering

Happy with the newly built CV joint, Caitland and Morgause devoured rocket ice lollies in the yacht club café.

We refitted the thrust bearing and CV joint. We were so proud of the engineering that it got a coat of best Hammerite paint to keep the stainless bolts clean and sparkling. It was a work of art and probably ten times the strength of the original.

Life on the water means diving on air tanks to re-fit the shaft anodes. The local dive boys refilled our tanks to 200 BAR. The commodore and his team were back onboard to witness the testing. Did we mention that the commodore was a diver in his other life? He was now retired but famed for his time in the water with Jacques — that's Jacques Cousteau to you and I. It's always inspiring how the heroes are there when you need them. Caitland teamed up for the sunshine photo.

CHAPTER THREE: *Entering The Roaring Forties*

The vibration was down, and with Polly Perkins sounding quieter, it was time to go, but nature didn't choose to match our timescales. The wind, as the forecasts had warned us, down here in Patagonia was changeable on a grand scale. You have to wait for a window and go.

Waiting for that window, the girls would be collected each day in the school RIBs for Optimist sailing lessons. They were taxied about the beach water as the children were

CNAS Commodore Porto Madryn diver & Jacque Cousteau chum

CNAS regatta chums

Optimist sailors

placed into their small sailing boats. The girls, proudly togged in the green club shirts, joined chums to hone their own skills on the wind.

There was time to explore the *Museo del Hombre y el Mar*, a social heritage from the ancient Indians of this land to the Welsh settlers and modern-day Patagonians. We also visited the two long high streets with a Carrefour supermarket and endless wool jumpers in the shops. There was a fine Malvinas/Falklands memorial and a town pier for the cruise ships who make this a home en route to Port Stanley and Cape Horn.

A yacht club dinner

We took a tourist daytrip for the far beaches toward Punto Loma and checked out the Ecocentre. Land of the sea and the history of the right whales, we walked through the mouth of a whale, a curtain of baleen teeth, through to the stomach of the beast. There were even sounds of digestion in this eerie place of learning. Beyond the whale was a saltwater pool of star fish, a viewing tower and a library with a children's room watched over by a giant soft fury teddy bear squid.

The sail school held a Saturday Regatta, where there was a barbecue in our honour, and by Monday, Caitland and Morgause went solo in their Optimist sailboats. The wind pinning us to the shore delighted the youngsters as they raced the waves with their dinghies like surfers on the crests of the beach.

Goodbye Puerto Madryn

With the weather turning by the day, the winds finally softened and came ready for a turn to take us south. The commodore instructed us not to leave before twelve, as everyone had organised a leaving regatta for us. The shore lines of Puerto Madryn

CHAPTER THREE: Entering The Roaring Forties

looked like the start line of a TT Race. In line formation and with their sails full, the Optimist school sailed towards us in fancy dress. Valeri, the coach, followed in her RIB, blowing her whistle, Hoquin was one of the best young sailors, he swapped sail rope for drum and was beating with all his might. Friends, parents and all the sail school staff wanted to bid us a fond farewell, and there were tears in our eyes.

The small sailing craft danced all around with the girls wearing their green club shirts. They were half black with ink where friends had signed them with goodbye messages we would remember forever.

South from the Golfo Nuevo

RIBs large and small circled the *Hollinsclough* in an acrobatic synchronised dance, whisking up white water like the skirt of an ice skater. With the team of boats crisscrossing our bow, sails and flags waving goodbye, the commodore's launch took centre stage, flying phonetic sign flags for "Good wind."

In the entire world, our hearts have never been as touched as by the kindness of the people of Puerto Madryn. We saluted with our colours, turned about with such sadness to leave, and set west across the Golfo Nuevo. We will remember Porto Madryn forever.

With fair weather, we took a short 35-mile leg across the Golfo in the afternoon sunshine to take a sheltered sleep in the small cove off Piramides, land of Dakar rally — a Star Wars desert of rolling cliffs. Small penguins guarded the edges of the beach.

Best-laid plans and all, by the early hours, the Patagonian wind had changed. It built to 30 knots, turned our anchor and pushed us towards the beach. There is no solution to a turning wind, moving tide and a shallowing depth at anchor. Why is it always two in the morning? We drew the anchor chain with the precision of the Florianopolis bridge — steady on the motor to match the chain pull holding us steady over the shallows. Shorter chain, more drag, less depth. That wonderful sound when the anchor clears the water and bangs into the stop, tonight the clunk was softened with ten buckets full of gooey grey mud.

Hold fast! Motor full on, and we escaped half a mile out into ten metres of depth — less protection but good holding. Sleep? Don't ask, but by 40 south experience of anchoring, you do get to sleep in the end.

The wind softened by morning only to awaken for an early lunch. We recovered on spam sandwiches, and by two, we were away again. Our adventure for the end of the world resumed — all south, 180 on the autopilot dial. The Ugrib weather forecasts were good. We celebrated with a bucket wash down of the bow to drive all that muddy gloop from the front to the back stern, around and about and over the side. There is something very relaxing about cleaning up the mess of a soft anchorage.

Back in the sailing department, we exited the protection of the Golfo Nuevo as we cleared the point of Punta Ninfas. The outgoing tide was not kind, as it drew three knots out of our speed. As if to say, "Wait, take one last look." This was a land of friends and memories.

Pushing down the Patagonian coast, we were approaching the Cabo San José. The evening winds soften, and the motor was on. That left electricity plentiful — water-maker whirring, fridge and freezer chilling.

Every evening was becoming colder. The night watch required full togs; it had been some time since the yellow trousers and sea jackets had been used in anger.

A giant full moon of suitable light made the sea ever more yellow, while toward the mainland, a warm orange glow of light from the city of Rawson at the head of the Chubut River filled the sky with a sign of civilisation. Tide runs and fast currents pushed us along for some good miles.

First kelp on deck

A steady morning on the motor gave us a late-lunch arrival off the Cabo San José point to run on the Polly Perkins motor against the wind into the bay of Santa Elena — 44 south. Totally deserted of humankind, this faraway circle of yellow rock hillside provided a splendid respite from the building Patagonian winds. It was our first kelp, endless strings networking out from the beach.

We were greeted by four boisterous penguins, and with the anchor laid in the shadow of Punta San Fulgencio, a pod of four dolphins came to check our chain and say hello. Golly, this was as far from mankind as we had yet been. There was a desolate beauty in the land of a seal colony, where airborne security was provided by the albatross.

Santa Elena blew 30-knot winds south-east as we ate our hearty breakfasts of cut bread toast and strawberry jam at anchor in a small gale. Feeling very much at home on the chain with only the noise of the water and wind as our companions, we found three evenings of tranquillity.

We were waiting for a favourable change of gusty blows for the next coastal leg south; it was a short run to Caleta Sara and Caleta Hornos. Our hopes faded away as a Valentine's Day passed without cards.

The wind came good for the morning of the 15th February, and we set a course all south direct for Puerto Deseado. Watching a wild red sunrise, we drew anchor of grey sandy mud from ten metres and had left our home in the bay of Santa Elena by seven in the morning. We made a flat sea exit from the most remote bay we had ever laid in. The first sign of life was a red fishing boat with nets down and an entourage of trailing gulls northbound as we moved out across the Bahia Camarones. Flat-topped mountains, ravines of red rock and wild yellow sandstone cliffs made for a magical view of this desolate place.

We came portside of the Isla Acre for the Cabo Bahias. We saw more breathtaking scenery of rock-scape oblivion as the Isla Rasa rocks came to our port side in a chicane run of coast and favourable tide races. Sunlight spread across the Isla Leones as we pulled of the coast into the large bay of the Golfo San Jorge. Blighty legends, George and the Dragon, the mighty lighthouse stood like Wolf Rock, Land's end.

Hollinsclough ran down the wind, her jib poled to port and the main wide to starboard, making space for Monopoly on the stern deck.

We had 30-knot winds to make eight and nine-knot ground speeds in the surf. Evening found us 45 south, halfway down the roaring forties.

The sky was a picture of apocalypse, lightning storms fit for Guy Fawkes celebrations and rainbows in the sunset with enormous fronts closing our starboard stern quarter for fire-breathing dragon monster weather. We were afraid to leave out our Corn Flakes for horror they would be toasted to Frosties in the bolt of a lightning strike.

It was a rough sleep with washing-machine spins in the window frames downstairs as the girls were pushed into the walls of their bedroom, with the mast turning the floor angles to match the sails.

A weather map formed in the sky.

The cockpit was a wild white-knuckle watch, but we made good speeds for an early arrival on the Cabo Tres Puntas. Sunlight rose to an escort of albatross, we had softer cliff faces to our starboard side, and it was a welcome relief to find clear blue sky with no scary bits.

We made a 30-mile dash down the low sand coastline, again absent of mankind. Beam-reach winds and eleven on the ground speed dial was rocket-ship pace to make an afternoon teatime arrival in Puerto Deseado.

Puerto Deseado was a fishing and commercial port void of landing space and pontoons. There was an enormous concrete wall where ships could side too for cargo and bulk loading.

There was no obvious yacht anchorage; our charts were outdated to a fishing boat takeover of the shallow water. Shrimping boats, rafted ten deep and stood about 20 in line almost closed the entry channel. With all luck a sleek grey *Prefectura Armada* clipper was in town and gave us a temporary berth on his side, ropes and fenders tied proper navy style to clipper GC70. The paperwork was sorted pronto.

CHAPTER THREE: Entering The Roaring Forties

Our stay with the Navy ended all too soon. *Prefectura* on duty at all times, they told us we had to move. We were advised to take the yacht mooring buoy, but for the life of us, we couldn't find it, only to realise it lay so close to the shore that we would be aground. Back to the GC70. "You just can't stay alongside us, but we have found you a home."

A tugboat birth in San Julián

Our new berth was stern to a giant tug alongside the pilot boat *Yamana*. We were delighted to have the security of the tide, but it was a shock to find out how we had to get off. Low tide is never helpful; we climbed the roof ladder of the tug, stood on its roof and a made a leap of faith onto a blue pylon for a rope ladder that took us up the side of the concrete dock to ground level. It's not easy for those of us who are scared of heights.

John Cogill, the sea patrol tug came to our distress. Their orange clad "tango crew" put the children on a captain's chair, held the rope tight, and by using laughing tones in Spanish, we managed to get across. They loved helping the damsels in distress.

Puerto Deseado was another Welsh settlement. This town existed not just for the port but the railroad that linked it to the north of the country. Steam had long gone but old carriages lay on display in the streets and a fine station museum told the story. To our delight, a bread shop and burger café was open.

At eight p.m., we were told *John Cogill* was leaving. The pilot boat *Yamana* we were moored against had to move and show the *John Cogill* the way out. We were asked to pull away, circle and return. That's a difficult call, but it seemed a good marker — time for us to leave. We watched the deep-sea tug move sideways from the dockside. The orange-tango-clad crew were "Thunderbirds are go" as they scampered about the decks. It was a precision manoeuvre for a 79-foot vessel in a tiny space. Exiting ahead of, the *John Cogill* left us proudly, head of a three-boat convoy as we all charged for the open sea. The AIS avoidance computer never stopped bleeping.

With a dark night and a long day, we made 48.5 south to Cabo Danoso, a giant pyramid of a mountain standing tall on a dark flat run of endless coastal rock front.

Sunset with no aeroplane trails

Bahia San Julián beckoned for shelter as it had done for Magellan Fitzroy and Darwin.

Whilst our GPS positioning was responding perfectly, the cartography maps were out by more than a mile. This was the first mismatch of any magnitude we had ever seen anywhere in the world.

It came as a shock. We were faithful to a computer system that had never let us down in the darkness of any port, and yet here we could see a huge abnormality before us. The San Julián estuary is almost a mile wide with sweeping shallows on the chart. We used the radar overlay on the chart system to re-match our base point, and Morgause took her regular duty on the depth gauge. Point two metres left… point one… "There's nothing on the dial! she screamed. We scuffed over a gravel bank, and she howled with joy. "Point two, point three, point four metres on the dial."

We edged in slowly, as if tiptoeing. The bravery and skill of Magellan, who came in without any charts, throwing lead and string lines for depth, was inspiring to the soul.

As we focused all our energy on the route, Commerson's dolphins played in the waves. They had strongly contrasting black and white markings like a Friesian cow, small fast and streamlined. We believe Commerson's dolphins are without any doubt the most beautiful dolphins in the world.

We faced up to the sight of a giant cross that stood where Magellan had held the first mass on South American soil in 1520. Five hundred years later, much of the area remains unchanged. Beautiful gravel islands roll softly out of the clear cold South Atlantic water to make fine homes for the Penguins. San Julián was a colourful picture of warm houses tucked deep into the estuary, the water calmed by a hundred twists and turns of shallow gravel banks.

We dropped the anchor chain a few hundred feet portside of a true-life replica of Magellan's galleon the *Nao Victoria*. It stood on the shore close to the town quay — what a sight from our bedroom in the waves. Living history as the anxiety of the shallow water faded into dreams, we could hear the crew of the *Victoria* shouting depths on their lead line.

CHAPTER THREE: *Entering The Roaring Forties*

We anchored next to Magellan

The *Prefectura Armada* office was waiting for our paperwork as we landed ashore in the RIB. A Mirage jet fighter outside the navy office honoured the Malvinas/Falklands War. Three English ship targets painted below the cockpit of this racy fighter plane were a testament to days gone by.

Golly, it was cold here. A warm chicken lunch at Naos Restaurant warmed our tummies where we checked out the tourist office pamphlets. Captain Fitzroy brought the *Beagle* here with Darwin aboard.

Back to Magellan, and all aboard for a fantastic history lesson on his ship *Nao Victoria*. Life in a Tudor galleon, waxwork figures and sounds of the sea for a history lesson alive with reality. The old people of the land had been huge men. Magellan called them the "Paton", stories of Big Foot, but they are long gone in all but the name of Patagonia.

A TV interview followed with local station UVC. Girls together, it was our very first time on the box. We faced questions about what we had seen on our travels, World Guide Day on the 22nd February and how the girls did schoolwork. We then dashed back to the town hotel to watch ourselves on the local channel TV.

Family together, we took tourist tickets for a run around the bay in a big red motorised RIB. The Pinocho Excursion team snuck in every gravel bank to come between the penguin colonies. Wooly hats, warm coats and tourist orange life jackets were very necessary.

Captain Morgause took the controls of the big red RIB from Ferdinand, the real captain. With looks of concern, Ferdinand turned to the tourists: "She is a world class sailor; we are in safe hands." Morgause pushed up the throttles, and the tourists hung on for their lives. She was a master of a RIB with more power than a fighter plane.

Argentinean Mirage fighter

49

Giant petrels and grey dolphin gulls filled the sky. The bay was strewn with islands covered by millions of grey pebbles. We could land and walk amongst the colonies of Magellan penguins. There were thousands and thousands of them, more than we would ever see anywhere else in the world. We stumbled about in awe of the numbers; if not for our orange life jackets, it would have been a world record number of sittings for black-tie dinner party.

Sailing with the tourists, Morgause takes the wheel.

San Julián was very sheltered, and gave us three nights of sleep that we cherished after so much roaring forties water, but as with this boat life, all these places come to an end. Diesel was low and so it was on Sunday 22nd that Fernando piloted us out of the bay in his smart red excursion RIB. The GPS chart mismatch was logged and emailed back to Blighty. The deep shores where the locals fished became a goodbye committee as everyone stood tall on the shore to wave and cheer as we passed by in our winter coats. We turned out into deeper coastal water and gave some more sea room to be sure of the chart match.

CHAPTER THREE: *Entering The Roaring Forties*

The 50th Parallel South

Acclimatised to the roaring forties, the screaming fifties confronted us. We made 50 south at 4:30 p.m. local on Sunday 22nd February, Girl Guide World Thinking Day, which was a cool marker for 50 south. We emailed blue mafia Girl Guide chums the world over. "50 south, world think day and thinking of you."

No pontoons in the ship dock

Days were getting shorter by the mile, and the evening went very dark for our arrival to motor into Puerto Santa Cruz. GPS and the chart match was spot on again now.

Dark it may have been, but the lights of the port shone bright on an enormous concrete dock. Tall concrete piers, very dark, not a ring or cleat to be seen, perish the thought we could climb it — no room at the inn!

We turned about the end pylon, took a closer look toward the gravel bank shoreline, and got a stroke of luck! The *Prefecture* Naval Clipper GC133 was moored on the inside of a tiny landing jetty. Their spotlight lit us up, "*Welcome welcome, pull alongside.*" A true joy filled our hearts; we were expecting to be turned away.

Life 50-plus south was a little more lenient. Good English, and there was more to come. "*We will check your ropes at 4 a.m. low tide.*"

It was time for bed in this eerie concrete jungle of towering shadows. One-thirty bedtime... who's watch was that changeover?

Early in the morning, *Prefecture* Officer Verdez came aboard to complete the transit paperwork. His English was good, and he informed us that "gasoil" and "agua" were available. We only had fumes left in our diesel tanks, but sadly, we were unable to organise diesel here. The minimum delivery was 35,000 litres. "We only do ships! You must take at least one tanker load for a delivery," we were told.

What a business getting onto that commercial quay. Puerto Deseado was nothing on this. It was 10,000 feet in the air up a ladder fit for Blackpool tower. *Hollinsclough* did look a picture hung onto the cutter all alone in this world of big docks.

The sun was shining upon the Jurassic slopes of Santa Cruz. It was 60 miles of coastline with tall, flat-topped layers of yellow sandstone just as Magellan and Darwin had seen all those years ago. Thoughts of the Magellan Straits were uppermost in our minds, and the urge to move on was strong. Go we must. After inform-

Commerson's dolphins – the most beautiful dolphins in all the world

ing the *Prefecture* of our departure, he returned to stamp our passports and wish us well. A parting gift of six locally caught fish were given to us — gutted and frozen. "Welcome, welcome!" the friendship of faraway places was very strong here.

Hollinsclough sailed smoothly away from the mighty docks of Puerto Santa Cruz, a family of seven Commerson's dolphins that lived in the bay came to play with us. Beautiful black and white little mammals with a temperament like Border collie dogs herding their ships. No doubt the most beautiful dolphins in the world, they are also some of the rarest. They loved to squeak and jump about by the bow with more energy than their larger Peale's brothers.

Heading back into the deeper waters, the Commerson's dolphins were scared away as huge Risso's dolphins as big as four metres long came to take a look at us. They were dark grey and big — very big.

Morgause called deeper water on the gauge, and we moved out to cross Bahia Grande. It wasn't long before yet another friend of the sea came and said hello. A South American fur seal leapt out of the water, twisting and turning in the air before snorting loudly.

Time and time again, he played around the boat in the screaming fifties. We had never seen a seal so active and never so far out to sea. We emailed our position off Puerto Coig for Roy on watch back home and knuckled down for a cold night.

We travelled 100 miles across the bay on good wind and saw miles of the tall, yellow, Argentinean, sandstone cliffs plunging into the ocean. The sunshine rose, the tide speed was building and the horizon was gone — what must have Magellan thought?

Cabo Virgenes lighthouse, marked with black and white zebra stripes, signposted the end of the cliff line.

Adding to the confusion, nature's finest creatures were on hand in strong numbers. Beautiful Commerson's dolphins were everywhere, penguins, albatross and

CHAPTER THREE: Entering The Roaring Forties

giant petrels were in abundance, and we had our first sighting of the slender-billed prion. They were all creatures fit for Noah and the Ark.

A tide race, the ocean draining away, the cliffs gone, and nothing to be seen ahead. If there was an end of the world, then this was it.

Magellan and his crew were brave men to sail this water for the first time.

With the comfort of GPS, we turned past the lighthouse beyond the end of the world and ran into 25 knots of wind on the bow. Well done the Polly Perkins diesel engine, just fumes left in the tank, but she wasn't going to let us down.

Punto Dungeness, a sort of dogleg of the world, marked the charted anchorage. Imagine dropping the hook off Wolf Rock, Land's End, and you are close to the vision. With tides here fit for the Channel Island race of Alderney, we had to hold fast for a run down the hill; it was no time to climb the mountain. This anchorage was set for the tide wait.

With long low banks of gravel shoreline, colonies of Penguins mapped the edges more accurately than the GPS. Anchor dropped, not sure of the gravel hold on the hook, we used every link of that 100 metres of chain. A buddy weight and a snatch rope were added to hold us in 30 knots of screaming fifties wind. The tide raced so strong that water ploughed down the bow as it ran beyond and off the edge of Magellan's world.

Anchored here at the end of the world, we lay as close as the chart could tell on the Argentinean/Chilean sea boarder line. For every VHF 14 Chilean call, an Argentinean VHF 16 rechecked our position. By the half and the hour, every 30 minutes, we logged our position back to Roy in Blighty.

We turned off the VHF radio sets and turned in for the night. If it was the end of the world, we needed to get a good night's sleep before we faced it!

Lighthouse at the end of the world

Chapter Four

The Straits of Magellan to the Pacific

Flat calm – 53 south at anchor Playa Prada Magellan.

"*Commerce and nature share the ocean.*"

"*10,000 miles from Blighty...*"

"*40 Knot gale force eight, that's Magellan for you!*"

Whether the world was round or not, we were sat on the edge of it. The Patagonian cliffs of Argentina plunged into the ocean, the tide raced by, and a bewildering array of species circled above, afloat and below. The light of the city attracting us was Punta Arenas, the southernmost city in the world, but it was history that surrounded our anchorage. We were in the footsteps of Sir Francis Drake, the first Englishman to circumnavigate the world. He took his eighteen-gun *Golden Hind* through the Straits of Magellan, which lay before us. His mission to plunder the Spanish ports of Pacific South America for Queen Elizabeth, he would return to Plymouth in 1580, so he was certainly not around to answer our question.

Joshua Slocum brought the *Spray* this way on the first solo world circumnavigation. Every day an adventure, it was to Slocum's book we looked for inspiration of anchorages. We followed his words for our trip into the Straits of Magellan.

Raising 100 metres of chain fit to save us slipping over the edge of the world, we left the gravel bank. An end-of-world home of 100 Penguin colonies in the open bay of this faraway place, the Punta Dungeness, 52 south, was left to our stern.

We turned on the VHF radio, Lima Four India called VHF 16 at nine a.m. local. The Argentinean *Armada* call was our last word from a country we had grown to love.

The red sea tug *Goldrina del Mar* lay to our starboard, we crossed the small bay to clear the lighthouse. "*Charley Bravo Mike Seven One on* VHF fourteen," nine a.m. local, February 25th for a Chilean welcome on the wireless. The next radio cue was the ship *Elisabeth Boy* bound out of the strait for Fox Bay. The Chilean Coastguard asked what nationality Fox Bay was. "Falklands," came the captain's reply. We were sad in our hearts to miss the Falklands, but following Magellan and Darwin on the *Beagle* down the coast run was a life memory we would treasure.

With 270 degrees on the dial, we were headed all west into Bahia Posesion, just north of the Bahia Lomas — how cool is that? This was land of gales; Slocum had been pasted so badly here that he climbed the mast to avoid a rogue tidal wave. The wind gods that had defeated our run to the Falklands were kind to *Hollinsclough* as we took a soft motor run into still air amongst the oil platforms. There were more than 30 yellow platforms of the fields SK, SN, DU and PN. A fat lime green helicopter made a close fly by to say hello as he moved about the platform landing sights before us.

There were more than a hundred rigs in the bay.

There must have been more than 100 yellow sculptures pass by our deck in the next hour.

Commerce and nature shared the ocean as a magnificent family of southern right whale rolled out of the smooth water in the morning sunshine — mother, papa and baby all together. Albatross and giant petrels filled the sky, penguins were everywhere, and more beautiful Commerson's dolphins guided our bow in the shallow waves of the tide race that favoured our 40-mile run to Punta Delgada.

This was our anchorage target. Hoping to get the best tide from the Pacific to the Atlantic, we invested great planning, keen to balance tide and distance travelled against water distance covered. Our sailing was born and bred in the Channel Islands — five knots forwards against six knots of tide the other way is no progress at all.

The water run at Punta Dungeness was hardly half a mile across, and a chain ferry shuttled from the mainland to the island of Tierra del Fuego. Tierra del Fuego was home to the ancient Indians of this land. We marvelled at the chain ferry vessel with both ends pointed with aggression fit for the greatest gales on Earth.

Scheduled to wait and jump a tide or two, we dropped the anchor, positively shallow at six metres. This was still Patagonia, so ten times the depth put 60 metres out with a balance weight and snatch lines. Two pilot boats sat to starboard, and another called the *Skua* moved towards an approaching ship. She was 100 feet long, and that's big for a pilot boat.

Summer weather it may be, but we had 30 knots of evening wind and a tide at eight knots. A wicked whirlpool of froth scurried around the channel, moving and shaping the gravel banks as it went about its business. A ship in repair lay dry on the gravel bank. We watched ships sail down the channel, their AIS destinations on the computer were a map of the world — east to the Pacific, west to the Atlantic.

Tide reckoning and a VHF radio chat with the pilot boat to be sure set our timescales for a midnight party.

In the middle of the darkness, amidst dreams of pontoons and electricity, we stood over the anchor winch. As the tide changed, we swung about, raising the anchor in mid-turn, and *Hollinsclough* smoothly set sail for a fast getaway with the strong Magellan water building to full flow. Lights bright enough for Oxford Street adorned the shoreline where tugs and pilot boats were moored for the night.

The night was black, and beyond the car park of pilot boats at Punto Dungeness, the world was void of all human elements. The water in our path was violet black, but it was fun to hear the spouting of Commerson's dolphins come to our side. Ghost riders, whooshing towards the boat with a high-pitched squeak of delight as, friends together, they took a jovial bounce out of the wave just before the bow. There was no one to see but the flowing ghost-like shapes. Straining our eyes into the darkness, they were black and white torpedoes making green flashes in the algae as they passed. Small lighthouses showed the way for the modern ship, but these ghost riders of nature were here for Magellan and now for us.

By lunchtime the most magic number came on the log. Morgause, depth gauge reader extraordinaire, pointed to the display. "One, zero, zero, zero, zero." We were 10,000 miles out of Blighty.

This was a wild landscape, 53 south in the Straits of Magellan. At first narrow, it was now wide as it shifted between small desolate islands. The sea teamed with gulls, albatross and petrel. *Hollinsclough* bagged a staggering eleven point seven knots, sail out and motor on to grab every moment's advantage of a wicked tide that ran all the way to the windswept shore of Punta Arenas.

Our addiction to city lights was quashed when we saw that this southern most city on Earth was little more than a sweep of buildings clung to the side of a rolling green mountain. There was a sort of oil pipe quay to starboard and a single commercial ship quay in the centre.

Radio VHF 14 bid a welcome to this remote commercial dock, but there was not another yacht in sight. All VHF here was in Spanish, so struggling with the language barrier, we negotiated a rope mooring onto the side of a ship. We tied side-on to a huge blue sea tug called the *Beagle*. This was not the *Beagle* Darwin had sailed in, this was the *Beagle* registered out of Valparaiso. It had famously towed in the Falklands missile victim *HMS Endurance*.

We were past the 53rd parallel south, so a Florianopolis spider web of rope was appropriate. We set triple cord lines with the wind blowing close to 30 knots on the beam into our side. With ropes tied, our discussions with the *Beagle* skipper was on our desperation for diesel. He had a small bowser truck alongside within the hour. We were unable to pay. No cards taken, only cash, and we had no currency for Chile, but we did find 200 US dollars for a top up.

Winds building, the port of Dover would have been closed as we watched a pilot boat come in backwards and moor in a narrow slot between the next ships. It wasn't like this in Europe.

The *Beagle* had a towing job and asked us to moved. Winds at 40 knots, we moved out like a local, anchored off for a couple of hours and then with winds down to around 25, we moved back into a slot on another sea tug. By evening, he was away and we were back to port side of the quay for a night on the small cargo ship *Rio Aysen*.

Breakfast brought a move to the sea tug *Skyring*. We took a morning walk to the bank for Chilean cash, and the port authority got a pipe to us for 1,200 litres of diesel.

Life in a commercial quay was an experience. We had three spider web moorings and one anchorage in 24 hours. Imagine mooring alongside the ferries in Dover and add winds mostly over 30 knots for manoeuvres.

The upside of commercial marina life is that no one complains about the sound of the generator. We had 220 volts and the trusty toaster in action breakfast, lunch and teatime. With a wild party raging on board, we would still be the quietest boat on the ropes.

Punta Arenas was also a big location for cruise ships, and we had 600 feet of liner vessel *Oceania Insignia* towering above us for the first of March.

Punta Arenas was all geared to tourists. We purchased penguin hats, postcards, badges, "Cape Horn/End of the World" and even a cruising certificate for the Straits of Magellan. The city lights were bright, with tourist stalls all around the smart City Plaza where a fine statue of Magellan stood tall to keep an eye on us. More Chilean cash from the bank machines and more diesel on board, we took the opportunity to get a five gallon drum of oil from the Copec Mobile petrol station and some big sacks of potatoes from the mall. We treated ourselves to a pepperoni pizza for tea — English abroad and all that.

The city was made up of concrete roads in grid patterns as square as a waffle with buildings rarely more than a few stories high. The wide, relatively empty streets were doused in sunlight one moment, snow the next, dry cold air, wind gusts and

CHAPTER FOUR: The Straits Of Magellan To The Pacific

everyone walking around all wrapped up. It was time for a woolly hat.

Many of the locals were of mixed origin, but the descendants of the native Indians were striking with their dark hair and wide faces. The Maggiorino Borgatello museum, with doors opened in 1893, displayed the contents of a nation gone. The Paton man, many sepia photos with curled edges showed the ancient Indians, naked with wild body paint and straight dark hair. Selk'nam, Kaweskar, Yámanas, and Aoniken natives were all displayed, including wigwams and a hunting life.

Tourists and working tugboats at one

Bones from a southern right whale big enough to make a climbing frame were on display. There was a modern floor of oil exploration and another of natural history where all the world was stuffed. Giant condors hung from the ceilings with lama and puma on the ground, and crocodiles, whale and shark in the water. Then there was a display of the Antarctic, including polar bear, penguin and albatross. It was a magic museum, all the more magic for its faraway location and relevance to everything around us.

Chilean paperwork is never easy, as documents needed to be collected, but the Port Authority is always open. Their smart flags of the nation and navy flew proud alongside a regional flag — blue sky, yellow mountains and five stars. We were welcomed by the officer of the day.

Like the Argentineans, the Chilean Navy would follow us closely via location email and VHF reporting eight and eight. Twice every 24 hours, we would log our location and double up the message to Roy keeping an eye on us back in England. The Chilean travel document for sailing is called a "Zarpe" and logged every step of the way, not to mention a few quid in charges for every location that's paid in US dollars or Chilean — it's a no-card world.

However, nothing is too much trouble: *"Call us at any time for any reason."* It was a welcoming thought for those long canal journeys ahead. But not before lunch. Toasties are toasties the world over, and a street café where the ham and cheese was melted on grills in large baps warmed us for the afternoon walk home in glistening snow that soon melted from the concrete roads.

Leaving the city lights of Punta Arenas, we were all south for an easy 30-mile leg to Bahia Mansa. So small it was scary, a few fishing boats all of them aground. We chose to move a few more miles south.

All the Magellan passage books are written for transit from west to east, so we read in reverse for our next anchorage target. There is the opportunity to make a loop run for Cape Horn, and we were sorely tempted for such a once-in-a-lifetime passage. But the season was slipping away, and we were set for a direct run to the Chilean Pacific coast. It wasn't many miles — how long could it take?

From Bahia Mansa, the big country scenery of Tierra del Fuego opened up with snow-capped mountains. It was a Snowdonia Lake District view on a whole new scale, windswept barren hillsides where only a few trees found ground to hang on for dear life.

Bahia Posesion opened around the next corner. *Hollinsclough* anchor went down in Puerto San Juan de la Posesion — 53.38.00S 70.55.50W. A replica of the first Chilean settlement on the Strait from 1843 lay on the hillside to our north. Fuerta Bulnes beckoned for the dinghy, a fine shore adventure amongst the big guns and wooden turrets. In 1846, all but three of the people here had died of starvation, and the old name of the bay was "Famine'! A cross stood on the far shore marking the grave of Captain Stokes. He was the first commander of the *Beagle*, and he charted the place before Captain Fitzroy's second trip with Darwin aboard. History was in every breath of steamy cold air.

The 54th Parallel South, Tuesday March 3rd

Soft cool evening winds mean early morning sails. Lifting the anchor in a flat stillness of water and mirror sky, *Hollinsclough* set sail for another day's adventure. We passed close to many fishing boats working the Straits of Magellan for Punta Arenas. A mesmerising mountain skyline towered above, and dark-etched shapes surging from the water's edge, each with caps of white snow. We drank steamy morning coffee to soften our frozen biscuits from the winds of Antarctica.

Chasing both tide and sunrise around Cabo San Isidro, the light above the mountains by Isla Nassau gave a glorious picture of 54 south as we closed Crown Point for a look proper at Cabo Froward.

Cabo Froward is inland of Cape Horn and Tierra del Fuego but is still the southernmost point of mainland South America. The top of Cabo Froward is marked with a large cross, the third of its kind. The first one was erected for the birth of the

CHAPTER FOUR: *The Straits Of Magellan To The Pacific*

Monsters climbed out of sea level.

new city of Punta Arenas. Twice the cross had been worn away and blown down, but with the visit of Pope John Paul II, the Patagonian Chileans wanted a new cross to be erected. 1968 saw a reinforced steel mesh design to beat the elements. Our view was with soft air and the blessing of the wind gods. This was a Rio statue designed for the wind in a trellis of spider web steel painted white. There was no climbing to the top of it like the one in Lisbon.

Beyond the cross, we made Cushing Point around ten. A tiny narrow exit for the San José River opened from a Himalayan-like range of mighty bare grey rock where a crescendo of water escaped into the straits. An equally apocalyptic wall of limestone followed, three or four miles of 1,000-feet vertical wall called Cape Holland. To the south lay glacial entries to the canals Barbara and Acwalisnan leading to the Beagle Channel and Drake's Passage not 100 miles from Cape Horn. Our hearts were sad to miss such adventure.

In the soft sunshine around 73 west, we spotted the French sail yacht *En Chemin de Cygne*. The girls' schoolwork went direct to languages and French for a chat on the VHF radio.

Bahia Fortescue lay close to starboard, one of the oldest charted anchorages on the strait,

The southernmost point of mainland South America

Fish for tea again!

mapped since the 1700s. French lessons over, it was time for history and back to Magellan, Darwin and Joshua Slocum. Shallow depths and we made a turn around Wigwam Island to drop the anchor in a small lake. Like Indians before us, we made a fire, but ours wasn't war signals, it was for marshmallows. We took photos for the Girl Guides.

On Wigwam Island, we began a new recycling process. With no port or marina with skips ahead, we had rubbish to deal with. We added all our burnable waste to the fire, and with some thrill watched it disappear. It was a great process we would follow for the rest of the world.

A small blue and white fishing boat, the *Sandra Vanessa* pulled in for shelter in the darkness. It is a great feeling when the locals choose the same spot. With big winds on the breakfast forecast, it was time for a morning cup of tea in bed — there's luxury. The *Sandra Vanessa* came alongside in the frozen morning air with three cheerful and chirpy fishermen. We swapped a bottle of beer for three huge fish. They were homeward bound to Punta Arenas.

Wind against us for the west, it was time to put out all the anchor chain and dig in for a few days of wind. The scenery was unbelievable, like Wales on steroids, 3,000-foot snow-capped peaks on every side, we could have been anchored in the top lake of Snowdon below the Crib Goch Ridge.

We had three days of fish for tea as the big winds built from a low pressure, forming a mighty gale around us. Regular 35-knot winds span 25 amps out of the wind generator for afternoon DVDs of the *Blue Planet* and Antarctic animal revision. Morgause tracked data for a line graph of the wind speed while Caitland focused on history with a look back into eternity for the Incas. We had time spare to clear the toilet hull exit pipes, like you do in a gale, and spring clean the engine. The thrust bearing and CV joint repairs were looking sound. A great pride fell upon us in this magical place of sailing history as we read the pages of Slocomb during the gale.

Building bravery to the winds, we snuck back out into the Straits of Magellan.

We left our campfire fun at Wigwam Island, Bahia Fortescue at sunrise on Saturday morning March 7th. Locked in for three days, we would make two anchorages in a day today. There was a little rain for the short run in the soft morning air with a fine escort of seals in our motor wake. Charles Island to port, then Monmouth Island before the wind unexpectedly picked up. We ran for the lee of Rupert Island as the gauges went red for a 40-knot gale force eight on the bow. That's Magellan for you.

The sea as white as the snow-capped mountains, we rolled around into Mussel Bay, came starboard of Dessant Rocks and found splendid protection for anchorage in a little Caleta below large trees on the northeast shore of the bay. It was a tremendous moment to find two southern right whales playing in the bay, blowing and tail bashing.

We had a cracking view beyond the whales of the straits, and the Ingles Passage outside went surprisingly flat post lunch. Tide and time waits for no man, so we upped sticks, raised the anchor and charged on for our second anchorage of the day.

We ran to the lee of Bonete Island, enormous rolling mountains of green trees signposting some shelter. Around Rowe Point and into Tilly Bay. It was no great distance, but every inch counts and flat motor runs down here seem the order of a good day. Wow! There was another yacht in Tilly Bay, *Nanoq* on a Swedish flag, tied in very tight to the shore. We hooted a few times, dinghy alongside, but there was no sign of life. It was clearly a nightshift boat, and all were fast asleep for a rest in the wind. In acres of kelp, we found our first sea otters. There were fabulous rainbows setting back into the surf of the open strait, but our saltwater-smeared windows did need a little rain.

Anchorage at Tilly Bay – 53.34.25S 72.24.27W

A Sunday morning sunrise run saw the anchor up in darkness. Small squid played in the underwater lights of *Hollinsclough* as she left the shelter of Tilly Bay in the Straits of Magellan. Out into Tortuos Passage, red and green port and starboard lights faded into the morning sunrise as we passed Bahia Borja with its famous sign boards of old ships anchored in a world gone by. We came about 300 degrees into the wide open space of Largo Passage, snow-capped mountains to the south, green Lake District peaks to the north. Seals splashed about and albatross left white jet wash beyond our bow.

After Carteret Island, we took sight of our first blue ice. It was breathtaking. We saw the snow-capped Mount Radford, but there was more to come. Layers of magical ruffled blue sheets all rolled up to the edge of the rock. By Abra Island, we could smell the Pacific, just 25 miles west down the Abra Canal. The cargo ship *Cop-*

pename, 360 feet long passed the other way bound for Lagos, Nigeria, as we turned hard starboard into the Peninsula Cordova for a nook in the mountains.

We anchored dead centre in a 200-metre circle of volcano rock wall at Caleta Playa Prada. It was as flat as a pancake in a frying pan, shelter from all sides and a view to die for. Rock ran to the sky on all windows for a truly alien-like anchorage of nature's most majestic splendour. Two tall waterfalls crashed down to the sea by our stern. The soft current moved our view a little as the hours went by like the viewing platform of a magic roundabout. The Swedish yacht *Nanoq* arrived, and it was time for English tea. It wasn't a nightshift boat but a single hander. We met Sverker Ullerstam, who was two years out of Sweden. No satellite for email, we asked how he got weather. "From yachts like you," he answered with a smile.

As steamer ducks rattled the water's edge, we made campfires ashore for a few days' fun on land. Woodlands and scrub to clamber about, Indian bread fungi grew all around. It rose from the dead wet stumps of old growth like mighty mushrooms. Red, tube-like flowers of waterfall plants hung in the sunshine between large green ferns.

The Magellan anchorages

We had time to explore both waterfalls and the west ridge rocks. The job of movement without paths limited our distance, but we scrambled about, making short pushes into the undergrowth. A reality of times old when people rarely travelled beyond their village became so clear.

Forecasts were favourable for the next steps, so we pulled the anchor up in darkness for a very early morning tide and better air to slip back to the channel. we passed Shelter Island into the brightness of a full moon. There were a few showers of rain to wash the yacht and darken the edges of the mountains. The straits opened up, we could see the marker light at Copper Point, and in a small cove lay an 85-foot vessel called the *Podu*, who lay at anchor amongst zebra stripes of black and white rock.

Tide turning, *Hollinsclough* made the mighty headland of Tamar Island to turn north proper, leave the Straits of Magellan and enter Canal Smyth.

We were in the Pacific for sure, and it was going to get warmer?

Chapter Five

Pacific to Puerto Montt, Chile

Seno, the blue glacier, 48 south.

"*The lighthouse was a prize to be taken.*"

"*Ice, majestic, beautiful and ancient, truly touching the soul.*"

"*Ahead lay a legend – Gulf of Pain (Gulfo de Penas).*"

Fairway Lighthouse stood on the island ahead. The Pacific, a new ocean lay before us.

We came about Fairway Island to a lighthouse that stood as a signpost to all modern sailors. It pointed back down the Straits of Magellan for the Atlantic Ocean behind us.

The lighthouse was a prize to be taken, and we motor-sailed into the shallows to rope *Hollinsclough* against the rocks of the lighthouse. Oil drum cleats built for the monthly supply vessels provided security. It was time for a very quick visit.

We walked the steep steps over a smart red metal bridge. A wooden helicopter platform greeted us, and then it was time for coffee with the lighthouse keeper and his wife. There were giant satellite dishes and two diesel generators, but the small lighthouse was almost toy-like — more of a communications platform really.

The water was crystal clear, and kelp danced from the rock bottom as we left the lighthouse to make for an evening anchorage. We were unable to find the *Armada* mooring buoy in the deep-water anchorage — maybe it had "gone with the wind." With the tide favourable, we were onward a little more north toward the sunshine.

The clarity of the long timescales people take to sail the canals of Patagonia was becoming clear in the wind changes. Becalmed to turn a mountain into a gale, narrow runs, endless stretches of kelp and rock edges unmarked, sailing was best left for daylight, and with short sunshine hours, that too was narrowed by the tide run times.

The pontoons get more solid.
Ropes loose for the tide at Fairway Lighthouse.

Our sunlight was gone as we made twelve more miles on the log beyond Fairway Lighthouse. We twisted through three shoals and past a shipwreck of Titanic proportions. A mighty rusty-red tanker lay propeller-side up above her watery grave of a rock bank in Canal Smyth.

Our spot for the evening was a narrow, difficult entry. With darkness upon us, the whole family was rock spotting through a ten-metre gap in the cliff wall. We were all the more focused after seeing that shipwreck. The cut was 20 metres across,

The graveyards in Canal Smyth focused our navigation to the metre.

real Jurassic Park stuff as it wound past two tiny islands and a long left turn that held back the view from our eyes.

Morgause had one metre left on the depth gauge, but the cut finally opened to the most beautiful cove of Caleta Darne. Sheltered still water inside turned to rings of moonlight in our turning circles. After making a depth check, down went the anchor.

We made 73 miles for the day, and that's a long way in the Straits of Magellan.

Caleta Darne 52.28.60S 73.35.50W

Sunlight greeted us with toast on the 220-volt generator and picture-postcard views. There was bush and green trees to every side; the contrast to the windswept rock was a wakeup call of escape. Green foliage climbed into mountain splendour for what must have been the most beautiful and remote anchorage of all our time afloat.

Friday the 13th of March. Never start out on Friday the 13th say the sailors of old! Our beautiful anchorage of green foliage was washed with rain for 40 hours. Friday the 13th turned our luck for sunshine. The boat was washed cleaned of saltwater as we left the enclosed bay of Caleta Darne and we ran out of the brightness into the shadow of 2,000 feet of Mount Filiu.

Kind wind and a good tide started the day at seven knots. A roller-coaster chicane run of buoy turns for Pollo Island tide station. Canal Smyth is different to Magellan, twisted with lower mountains, almost hillsides, and winds moved rather than funnelled.

CHAPTER FIVE: *Pacific To Puerto Montt, Chile*

We bagged eight knots across Guacolda Bay in a rush of wind, motor-sailing with good sail on the boom into Fortuna Bay for a look at 6,000 feet of Mount Burnley playing hide and seek in the clouds as we closed Cutter Island.

By Rennell Island, we had lost the tide and snuggled into the lee of the port side snow-capped mountains that grew with every mile forward. We had become quite expert at judging tide run. Pacific swell drove the ocean sideways into different sections of the canal, and it had to be judged like an estuary flooding rather than a tide rising. The best line was picked like a measure of the TT motorbikes back where we started our voyage on the Isle of Man.

A small white grave cross stood on a tiny rock islet of the shore. Who was or what was buried there? The steepness of the sides brought tumbling waterfalls. They repeated almost every 100 metres as if the hills were weeping for those who passed here.

Leaving Canal Smyth and taking a starboard turn into Victoria Passage, we saw a small red fishing boat working his lines. Hunter Island and Penla Zach for Canal Union, each turn was a gamble for better tide and wind on our Friday the 13th lottery that brought no bad luck. Shivering shinny mountains of snowcaps and blue ice beamed above. We found a radar reflector pole on a lone rock poking out of 2,000 feet of depth. It stood like a single flamingo on one leg for the betterment of man. Farquhar Passage for the last leg, wind moved against us, and we ran to lee for the shallows of 100 metres.

Mountains in the sky

Untouched anchorages

We made a full turn into Canal Sarmiento and the bay of Abra Lecky's Reach. Another untouched anchorage in the circular cove Caleta de Balandra. Four metres deep in a spaghetti of kelp meant hardly 30 metres of chain down to the anchor as we had no more than 50 metres to swing in this tiny haven.

In total darkness, we switched the underwater lights on to reveal the most bazaar view of the kelp forest. It swung about gently below us in an eerie forest pattern. A stillness in the wave like movement and clarity in the water fought the evening dreams that we were at one with the deep forest of ocean.

During the morning exit, the anchor drew enough kelp to cover the deck from bow to mid-ships as the hydraulic windlass strained at the chain.

As we cleared the kelp, 1500-foot mountains ran north up the Island of Piazza for seven miles in a grey sunrise morning on the Canal Sarmiento. Drawing parallel with Cape San Vincent 51.2 south, we could see Easter Island.

Well almost, portside lay a view of the mighty Pacific proper as we skipped across a gap in the mountains. No beautiful blue paradise but a big grey monster with teeth that spat swell. Good protection then another small gap at Sharpes Passage, and it looked scary out there.

Full colours for lunch, and the excitement of the day. Passenger cruise liner *Insignia* was southbound with 300 passengers to wave. Close to, green to red, Morgause on the trumpet, and Caitland on the pink hooter. After chatting on the VHF radio about glacier visits and bread baking, we put the cutter sail out for the photo — tourist postcards to Puerto Montt please!

Closing the head of Canal Sarmiento, we came too starboard side into a small open bay marked by a channel light on Pounds Island — another haven of tranquillity in green hillside protection. King Crab reported in the bay were elusive, but there was time for a dinghy trip up to the white rapid waterfall. We tied up and took small animal trails through the bush for a view of the upper lake and a campfire to burn rubbish and eat marshmallows. You don't get campfires on cruise ships!

CHAPTER FIVE: *Pacific To Puerto Montt, Chile*

A Sunday morning sail through the La Piedra passage put Canal Sarmiento behind us. *Hollinsclough* was in the heart of the mighty Chilean fjords as she ran east into Estero Peel. Four large dolphins were on hand for a greeting party. There had been no sign of Penguins since Fairway Lighthouse leaving the Straits of Magellan, but we did find our first small icebergs here.

A small red passenger liner *Skorpios III*, 230 feet with 60 guests, gave us a hoot on the horn as we came about to pass narrows between Orrego Bay and Zapta Island. Where do these names come from? We had entered Heusser Passage, where the water turned bright turquoise and bubbles around the bow sparkled in freezing spray. The prize ahead was Amalia Glacier.

Amalia Glacier

Amalia Glacier was the first glacier we could sail to the face of. Terrific, white and snow-topped, the twisting base stood in the distance; ruffled blue ice came down and touched the sea ahead of us. Broken mini icebergs bobbed about as we closed the ice wall.

We can put this in the parents' gin.

Speed down to dead slow ahead, we munched slowly forward, crunchy, munchy crackles around the bow. It was totally wicked to be in melt water from the time of the dinosaurs.

"Take some home!" shouted the girls.

We put poles out to collect small chunks of ice debris.

Not a glacier was to be missed after this.

Hollinsclough sat stationary below the growling glacier, blue light reflecting from the crystal world of ice water dancing all around us. Dinghy down, it was time to get really close.

Back aboard we opened the flag locker, chose the newest and cleanest flag of Royal Chanel Islands Yacht Club we had and ran it up the jack pole for a post card home.

Ice, majestic, beautiful and ancient, truly touched the soul. We were addicted to the magic from the moment we entered that fjord, and beyond Amalia, there was not a glacier to be missed.

Our evening anchorage lay around the corner from Amalia in Estero Peel. We savoured the time on the ice face, only leaving as the sun began to touch the mountaintops.

Safe from floating ice a little west of Fritis Point, we passed through a rock wall narrows of 20-metre width. In the stream of water through the gap, the girls held fend-

Blue for the ice and the flag of the Royal Chanel Islands Yacht Club

CHAPTER FIVE: *Pacific To Puerto Montt, Chile*

ers in case of a bump. Inside, we found Caleta Villarrica on the Wilcock Peninsula. Towering rock walls like Cheddar Gorge circled us like the walls of a Welsh castle guarding from the weather.

Our catch of glacier ice had melted to water in the bathroom basins, but it made for a chilly, exhilarating wash before tea time. Drinks — chin chin — chilled with more of the magic frozen crystal. How steamy the mirrors turned as ice melt fought the warm air of the cabins.

The Rain Zone

Estero Peel was almost the last of our screaming fifties. Landfall rain from the Pacific made mornings grey and wet. The mountains of the Chilean fjords guarded us from the big swells of the mighty ocean, but this was the rain zone and sunshine had become a distant memory.

Another red passenger cargo ferry passing by cheered us up. The *Evangelistas* was bound for Puerto Natales, 400 feet long in the morning mist.

Hollinsclough turned north, 300 on the autopilot dial for Canal Pitt and the northern narrows. The run of water was around a half a mile wide and 30 miles long.

The tide began with us and later turned against us. The wind all about, the weather was an entirely local phenomena, but true to form, it was raining. Ah, but for the unbearable heat of Rio!

The wind turned harder against us and funnelled over 20 knots, making it a long, slow slog of the 30 miles. We took a breather for a quick look in the caleta by Chatham Island. Small waterfalls and an otter pool were tucked into the rocky green sides that held back the wind.

There was sunshine to be had, so onward north against wind and tide on the strength of Polly Perkins. We moved between Profirio Island, its narrow rock ledges closing our route up the channel to a third of a mile. The day's fight with the wind and tide over, we made the junction of Expectation Bay to port and Sarratea Bay to starboard.

Snapper Caitland on bow duty

Entry gaps are getting narrower.

Caitland had taken stance on the bow and was senior Patagonian photographer by this point.

Galaza Point focused the autopilot to a narrow gateway at Seno Fuentes. The 20 knots of grey wind whipping down fell away to the mirror calm long pencil-shaped bay of Caleta Finte at 50.21.80S 74.22.50W.

Tall green walls rolled upwards into steep mountains, blocking the best of the late sunshine. Our eyes were drawn to the water, which was awash with colour. River and waterfall tumbling from the mountainsides made this joy of colour below us fresh water without salt.

With steamy breath, we peered in to see swarms of giant jellyfish rising from the bottom all about.

There were pastels of pink, red and orange as four giant frilly legs finished in bells as much as half a metre across. The place was alive. We switched on the underwater lights for an evening of Blackpool illuminations. It was a natural phenomenon of grand proportions.

We saw more swarms of jellyfish in the morning sunrise. Raising the anchor chain, the ten-mil links collected them in oily colours. As they lay on the bow, they withdrew to a smear of their former selves. I put gloves on to delicately put them back as we left the marvel of this jellyfish land.

CHAPTER FIVE: Pacific To Puerto Montt, Chile

Back around Galaza Point at nine a.m. on March 17th, Polly Perkins was at full steam against the opposing current and wind.

Kentish Island to port, and we were into Schroeders Passage 50.23.00S 74.25.50W.

We set for Seno Tres Cerros Channel, which was unsounded but well reported. Man had been to the moon, but no one has ever mapped the depths of this passage. The shallowest run closed to some fifteen-metre depths on this water without a map. The very green mountains had tears of rain making streaky waterfalls as the vista opened into the top of Canal Concepcion. We sighted the tanker *Cap Finisterre*, 650 feet long and bound for Bahia Blanca. He wished us well on the VHF radio and recommended the sea food in the fishing village of Porto Edén.

We turned north into Esto Lecky and finally caught some wind to turn off the motor and take six knots on the sail. We had the cutter full and a third of the main to sail on the wind. With so much time fighting current and wind turns in the narrow canals, it seemed strange to be on the wind.

More strange was life almost in a traffic jam — two ships in a day! The *Hergen*, 500 feet, southbound for Cabo Negro, the AIS locked in and VHF was on to chat. "Wow! Ten thousand miles out of England. Good luck and good winds."

The next big number for our dial was 50 south. Returning to the roaring forties, we left the screaming fifties behind at three p.m. local in the softest wind you could wish for on a flat sea with sunlight dancing on a slow

Artwork and music to match the maps

current. Wild seals hunting for fish acted as an escort. We blew hooters and celebrated with a colourful location banner — 50, a work of art. How a manmade mark on the Earth gives us a new momentum.

Our evening anchorage was aptly named Caleta Refuge. Well protected with thick wooded shores on Wellington Island, we anchored bow to the waterfall set into wind by the otter pools and found it had fabulous acoustics for reverberant echoes to shouts of "Boo!"

Caleta Refuge – 49.52.60S 74.25.00W

Two ships for a traffic jam, and sunset made it three. The 400-foot Russian Antarctic explorer vessel *Yuzhmorgeologiya* passed by our anchorage as we settled for the evening. In the best English on the VHF radio so far, we used his call sign of 9VNV74 instead of his name! He forwarded our position to the *Armada*, gave us a brief weather forecast and commented on what fun it was for us to have the whole family aboard. Cold War? James Bond would have checked it for a spy ship. Jaws waving goodbye from Rio smiled and let it go by.

Leaving Caleta Refuge in slack low water tide, we headed all north to break the sunrise. We looked into Caleta Sandy on the way. An alternative anchorage, it was very similar to Refuge. We made Labouchere Point for a nine o'clock breakfast.

At Averell point, we committed to the one-way streets of Saumarez Island. The Chilean Navy only permits this route in daylight. It is so narrow that it's a one-way street north into Pardo Passage for Canal Escape. Just to be sure, six Peale's dolphins joined our bow for the one-way trip that had tourist views to die for.

Three thousand feet sides rose stepladder fashion from sea level to the sky for three miles. The narrowest moment was Passage Piloto Pardo — Cheddar Gorge but in a boat. When the clouds finally let go of the mountaintops, the sunlight poured in and the water ran wide, returning to the more conventional route of Canal Grapler.

To some surprise, we had been in the lee of the mountains without a wind funnel. Back in the open, 30 knots of wind came down upon us with sea breaking over the bow and a ground speed below four knots. We could almost smell the fishing village of Puerto Edén. It was the first civilisation in 400 miles and most of a month, but it was beyond the grasp of the sunlight.

We pulled into an age-old mussel smoking camp. It was long abandoned but a haven of green trees and soft pink floor of shells. Shells, shells and more shells from the water's edge to the forest. We dropped anchor, tied some side ropes to the near shore and set about an exploration of the shell beach.

Caleta Moris Stella – 49.22.36S 74.25.05W

Dingy ashore, our feet danced on the crackle of shells for a campfire cindering in the history of the old smoke fires of the mussel fishermen. The smell of the mussel farms was all about us for bedtime.

CHAPTER FIVE: Pacific To Puerto Montt, Chile

A fresh sunshine 49 south day for Passage del Idino, and a final 20 miles to our dream of the lights of the city. Wind still against us and motor on, we passed portside of Grossover Island for more lee and then edged up close as you like to the rocks for the last of the run into the bay of Puerto Edén. It had been 450 miles and a month with no sign of a building.

Smoked mussel shells under our feet

A large green *Armada* buoy awaited, but it looked very worn. We came alongside the small patrol boat tied on a boarding platform and were asked to anchor off. We dropped the chain and took to the RIB dinghy to report in with our zarpe travel papers. We had a full update on the computer from all the satellite logs ahead of us, shipshape, dots on Is and Ts crossed, but our clocks were out by an hour for winter time.

To the horror of the *Armada*, *Hollinsclough* was gently dragging her anchor towards the open bay. We used their concern to speed the stamping of papers. These documents would clear our travels through Chile to the real city of Puerto Montt. By standards of the Magellan anchorages, we had acres of space to drag double over.

The girls had become expert in the dinghy at spacing rope ties to the shore, and our ability to get that anchor out of kelp and jellyfish was perfected second to none. There was still time for an Armada photo.

With smiles of salty sea dog experience on our faces, the *Armada* team hurriedly clarified a move to the Puerto Edén town quay. City lights!

Need this photo to be quick, as the yacht's going without us.

Puerto Edén – 49.07.60S 74.24.80W

The town quay had been refurbished to brand new only three months earlier. *Hollinsclough* looked a treat alongside smart red ladders and tyre fenders with shiny new chains. It had floodlights running on solar panels and wind generators just like our own.

This was not quite a city, as there wasn't a road in the town. The town was made up of maybe 100 buildings, mostly of tin with colourful paint. What you may call the roads were paths, and all of these were boardwalks built to survive the rain. Everything arrives by boat. Can you imagine a town without roads — that means no cars, no motorbikes no bicycles?

Puerto Edén was a land of the original indigenous Indians. Three marker boards on the boardwalk trail explained the age-old shell fishing we had become familiar with in the anchorages. Thirty per cent Mapuche, nine per cent Kaweskar Indian made up the population.

The Chilean government has made a strong commitment to supporting such populations. Amongst the tin houses stood the school. Like the jetty, it was very new. The school made their broadband connection available to us, and the girls attended art class for two afternoons. The art teacher spoke excellent English. Pastels and line drawing need no interpreter, but the school environment was a welcome return of young chums to the girls' remote life in the Patagonian canals.

Some small shops for potatoes, fresh eggs and bread helped our stores, and we put our order in for what we believed was four loaves. What looked like the post office was a small surgery; the Forestry Commission had a bigger office. We purchased two large blue barrels of diesel, and the policeman helped siphon them from the quay down into our tanks.

We returned to the bread shop, they had baked not four loaves, but four *kilos* of bread for us — it's sandwiches for a week then!

Puerto Edén — the most content and extraordinary faraway town you could ever come by.

Sad to leave Puerto Edén, we pulled out into a good forecast. Our diesel was good, and our freshwater tanks were full to bursting. It was a late morning start to get a favourable tide. Ahead lay the narrow tide race of Angostura Inglesa.

We moved starboard of the little island in Media Canal for the very best water. Four hundred and seventy feet of cargo ship, the *Cholera Narre*, registered in Bangkok, came by in the strength of the tide and said hello on the VHF radio. Bound like us for Puerto Montt, but he soon ran into the distance as 35-knot winds came out of the widening Messier Canal. In wide-open space and down to two knots, we threw in the towel and turned in behind Vitoria Island. In the large trees we found a tiny nook and slid in close to, touching all sides of the banking.

Vitoria Island – 48.54.17S 74.21.75W

With a big team effort, the dinghy was down and we had four ropes out to hold us fast in the trees. You could almost step ashore in ten metres of crystal-clear water. So close to shore, we were very protected from the wind. If there had been civilisation, we would have been in someone's back yard, but not a day out of Puerto Edén and we were back in the void of the Patagonian ice fjords of southern Chile. There wasn't a road for hundreds of miles, and the sky was empty of jet trails. Parked up in the wonder of the wilderness is what makes this place so far, far away. The larder was stocked full with bread and eggs, so it was scrambled egg on toast for tea.

The Patagonian winds dropped in the early hours of Sunday 22nd March, so we made for an escape from Vitoria Island at sunrise. Golly, the ropes came away easier than going on. Our four-point tie away, we left for calm water and almost seven knots speed heading north in the Messier Canal. There was a good view of the mighty shipwreck of Bo Cotopaxi sitting in the morning sunlight. We snuck starboard of Williams Island for even more tide speed across Tribune Bay. An old chum called on the VHF radio. Passenger ship *Evangelistas*, who we had last seen at the Amalia Glacier, called to forward our safe position to the *Armada*. Well done them.

Seno Glacier – 48.43.50S 73.58.00W

Our meetings with the *Evangelistas* must mark glacier visits. We turned out of Tribune Bay from blue water into milky white wash as we headed for our next prize, an adventure twelve miles up into the head of the Seno Glacier. Our ice addiction was in for a treat.

We were graced with flat water and wide open sky of bright sunshine. What a joy to get weather this good for the big beast of dinosaur ice. The blue white wonder of Ice-Age grandeur wound its way down from the horizon. We passed through a great deal of melt water and made two more miles to the face. Bergs carved before our eyes in thunderous crashes — nature at its mightiest. The ice was so blue Dulux

CERTAIN DEATH IN THE ICE

would have painted it Mallard. Strata lines of millennia were marked out clear as jam in a Victoria sponge cake. Not a road, airport or tourist in sight.

Hollinsclough pushed up until the ice was tight around her. Then we lay silent, motor off and time for lunch as we silently drifted back down the ice pack.

As if it was time for bed and our ice excursion was over, the lights went out. A mist came down from the enormous mountainsides, and it was time to leave this fantastic place.

Caitland and Morgause on their very own iceberg

Caleta Ivonne – 48.39.85S 74.16.30W

Opening back into the main canal, the water parted, and blue Pacific and white ice melt contrasted with a separation as clear as a line on a map.

Bedtime proper lay a little north in Caleta Ivonne north of Launch Island in the Messier Canal. Ice from the Seno Glacier

Life without an anchor

loaded up the boat's buckets, basins and sink as it had done at Amalia. The ice twinkled with frozen magic in the evening light where we lay safe. We had no anchor down; the girls had put a two-point rope tie across the tiny bay that made a fine Patagonian home for the evening.

The windy morning of Monday 23rd softened for a fine day northbound continuing up the Messier Canal. Warmer by the mile, Middle Island was surprisingly in the middle of the canal! It was a large green monster, but the sides all around the mountains had turned much greener as well. We could have been on a Derbyshire walk down Dove Dale, just on a larger scale. Lizard Island marked our anchorage closing 48 south. We turned in short of Fleurais Point north of the tiny Phipps Island for a sheltered bay with a wide roaring waterfall at its head.

Puerto Island – 48.03.10S 74.35.35W

There was mention of an *Armada* buoy, but no sign of it appeared. We lay to a swinging anchor set well in the current of the freshwater waterfall. Puerto Island was on Swett Peninsula. Using the dinghy for a look ashore, we found the chain remnants of the old *Armada* buoy.

Stillness and sunshine

Stillness and sunshine set in the waterfall north of the glacier fields brought a great contentment to our windy travels as we tucked into more of the Puerto Edén four kilos of bread.

A fish dinner served on deck complemented the bread feast; the canopy sides rolled away, all things relative, we felt like we had closed on the equator.

Gulf of Pain

Ahead lay a legend. The price to pay for life in the canals is a return to the ocean swell in a bay named by its reputation. The bay's name is the Golfo de Penas, which translates to "Gulf of Pain".

Tucked up in bed till almost nine in the warm morning air, our minds avoided the reality of the Gulf of Pain ahead. Warmth brings rain, and there was lots of it. A good look at the weather forecasts set our minds to leave our tranquil anchorage around two. We would make a direct run through a soft wind night to clear the open Pacific across the exposed gulf the following day.

The 500-foot tanker *Kiwi Spirit* passed our anchorage around twelve; she was bound in the opposite direction for Punta Arenas, where our canal adventures had begun.

It was time to tidy up and set the boat for sea. We strapped down everything that moved, and wedged the rest with teddy bears, pillows and duvets. Life may have been windy in the canals, but it had been mostly level.

We stepped forward into the Pacific Ocean, the far side of the world, with vigour. Hold fast!

The tide run was unexpectedly very strong. We reached San Pedro Lighthouse much later than expected at seven p.m. March 24th, and the land and sunlight began to slip away. We were 330 degrees into the swell of the Pacific. The wind was, as ever, higher than expected. The rolling of the sea was upon us, and it was a good job we had chewed the bread well. We had been in the canals so long that we had forgotten rolling. At eight p.m., we made a VHF position call to the *Armada* station at San Pedro Lighthouse — 47.40.00S 74.50.40W.

Sixty miles across the bay at five knots predicted a twelve-hour arrival for eight a.m. local at the VHF radio call point of Cabo Rapier on the other side.

The Golf of Pain gets its reputation for being shallow water. The Pacific swell breaking in turns and reflects from the wide 60-mile half circle shape. The reflection of the swell return tears it up into a rough ride in the best of wind.

CHAPTER FIVE: Pacific To Puerto Montt, Chile

During our crossing, the winds grew in the darkest darkness you ever saw. The waves breaking over the bow began clearing the canopy roof for a wild but bearable dash. The good news was that the building winds brought us in two hours early for a six a.m. sunrise. We made the VHF radio call to log our position as 46.39.00S 75.42.00W and change the watch.

Tide circling down the wind in the Gulf of Pain.

The mountains of Cabo Rapier were not disappointing of any world headland, true granny teeth specials all pushing 3,000 feet for almost ten miles of golden brown rock-scape. The vista was fit to stabilise the stomach with awe at what force is needed to stand up to the open spaces of the Southern Pacific.

Clearing the tormented reflection of water from the Gulf of Pain, the mountains of swell became so high they broke the view. The swell stabilised into rolling hillsides. They were so long that we rode them softly without crashing out. We climbed up to see the surroundings, and then ran down again, returning into the ocean.

Turning the autopilot in a little at a time brought good course and began to push us into the seven-knot territory on the sails. The strong tide slowing our entry into the Golf of Pain had taken its toll and not repeated its surge on the exit. The target to return to the shelter of the northern canals was Canal Darwin, but we chose to divert and make an easy evening arrival at Skyring Peninsular shelter of Seno Pico Paico.

Losing sea room to run back into shore, the black ocean became turquoise. It crested white froth where it met the stone shoreline. The swell was broken in anger. We rode a roller-coaster funnel of the froth to enter the Skyring Peninsula in another breathtaking arena of mountain-scape at 45.58.10S 74.59.58W.

Turning the last horseshoe of rock mountain, the lake of Pico Paico went shallow to a gravel shore that drained the energy of the swell, and a magical mirror calm came about the water. We dropped anchor and enjoyed the sleep we had dearly earned in the swell of the Golfo de Penas.

It was no time to remove the teddy bear packing, and we left the Skyring Peninsula anchorage with sunrise on March 26th.

Like a coliseum of nature with mighty volcano mountains on every side, we ran back into the open swell of the Pacific Ocean. As we peered into the ocean, it had turned orange. What looked like a million gold fish rose up from the violet crests of water.

All the fun of the fair — roll up, roll up, gold fish on the starboard bow! It was red krill and it swarmed everywhere, thousands and thousands of them. David Attenborough and the Blue Planet team would have been proud of our spotting. We tried to catch some krill in the fishing nets, but running seven knots it just tore the net out. There was no prize today.

Lunchtime brought another giant as we sighted the 1,000-foot passenger liner *Radiance of the Seas*. We chatted to the only English officer aboard on the VHF radio. Their film of the week was *Iron Man*; it was an up-to-date contrast to *Hollinsclough*, as we had a season of a Jeeves and Wooster DVD set. *Radiance* had 1,500 guests aboard, and she could take 2,500. That would have devoured our four kilos of bread. She logged our position for the *Armada* on the approaches to Canal Darwin as 45.28.70S 75.49.30W.

We had sampled a return to the open Pacific but were grateful for softer water as the mountains rolled back up the side of our world. Canal Darwin, the main route through to the middle Andes, was before us.

Twelve hours of motor-sailing put our anchor down in Puerto Yates, three miles into the tiny Canal Williams on the side of Isla Garrido. Every spot was a little greener, and it was our first day on deck without the big boaty coats on — sunbathing was close! The mountains were huge, but the size of it seemed very open to us. The passage books reported it as a fine anchorage, but it did just seem so open!

The sea temperature here was way up the dial, making fifteen degrees. Little black and white chaps bobbed about, not Magellan penguins but their close relatives the Humboldts. We had to get the big book out to identify these little blighters. Can you believe we would hand-feed the Humboldts back in Blighty when we found them in the Aviary Zoo of Harewood House near Leeds?

Not to forget we were still deep in the roaring forties, an evening of big winds in the sheltered bay of Puerto Yates was going to be no picnic. In the early hours of the morning, the boat swung from side to side, throwing us all out of bed. Then just to be sure, we were awakened by a loud bang! Don't all Saturdays begin with a bang?

The shackle attaching the snatch rope to the chain had broken. It was a big enough shackle to believe it could never break. The small safety shackle lodging the chain fell to the deck and broke before our eyes as we looked at the situation. The weight of the boat was pulled direct to the windlass, and it groaned with the strain, but it held steady. It was a team effort for anchor rescue, and the girls kicked in the electric motor, pushing the hydraulic pressure up. With a force-six average on the wind gauge, the gusts were up in the big scary 50 numbers.

Canal Darwin

We used the windlass to pull some chain links out of the strain and re-attached a new snatch rope with the beefiest shackle that would pass the ten-mil chain.

It was a long wet morning as a big wind battered down upon us. We were protected from the might of the roaring forties Pacific gale outside, but this place was far more open than the passage books gave it credit for.

The wind dropped and turned, and the move lifted our anchor and let us drag into the bay. Time to leave then? There was still thirties on the wind dial, but a few miles back into the Darwin Canal, and it was on our stern. Swell as high as you like, a running tide and 30 knots of wind gave us a roller-coaster run of ten knots ground speed with six feet of mainsail.

Every mile up the Darwin Canal shallowed and subdued the swell. We grinned and enjoyed travelling the crest of a wave powerboat style — what a change to have direction in our favour.

A dozen miles up, one of the small islands made a fabulous lee from the wind. Sails down, we had the motor in reverse to slow us up for a tight starboard turn entry. The anchorage of Isla Marcacci brought sea lions, otters, and rubbish. Our position was logged as 45.24.30S 74.09.35W.

We dropped a two-point rope to tie stern and bow and used the dinghy to run the anchor back out on the chain for a fine three-point mooring. We were ready for more wind, but it was the afternoon sunshine that came to say hello. The water was calm, and life was back to normal for bedtime with our new sea lion chums. It was so easy to sleep in the security of rope after our excitement in Darwin but such a shock to find common rubbish — plastic bags and pop bottles bobbing about the edges of this sanctuary of eddies and tide.

La Darsena – 45.00.70S 73.42.50W

Sunrise and back into the Darwin Canal. We had a good wind from the stern, so we were set for a 40-mile sunshine and showers run to Dar Island. We needed very little autopilot, as the wind was bang on the stern. The top score for nature was a 38-mile run, hand steering, hardly blowing — old sea dogs or what?

Our first sight of the Chilean salmon fish farms came beyond Negro Island. We spotted the crane boat *Redes Sur II* and then the *Coho* cargo boat at Mairo Island. We were all north leaving Darwin Canal about eleven and found our chum the good ship *Evangelistas* in Canal Moraleda. Green to green as we hugged the shore for best wind.

We made an afternoon arrival south of Dar Island for a gap between Isla Tangbac. Beyond a low sand bar, we found a small fish farm camp and passed to find a wide sheltered anchorage in La Darsena.

Filomena Island – 44.26.50S 73.34.30W

No roast beef and two veg for Sunday 29th March, but we had another good day on the wind with a favourable tide. As the canals widened like small lakes, we were making far greater use of the sail in our push north for Puerto Montt. We sighted the Navimag ferry *Puerto Edén* at eleven around 44-50 south, and he took our position for relay to the *Armada*.

Our evening anchorage was Filomena Island. A tight space between a small islet on the north east side was beautifully sheltered. Woodland touched the water between gaps of golden red rock. The fishing boat *Starbuck II* drew into the anchorage to advise us on the very best spot for depth.

There is no better local advice than a fisherman. Safely anchored and set to his best advice, he turned about and headed back out for an

The local fisherman told us to anchor just here.

CHAPTER FIVE: *Pacific To Puerto Montt, Chile*

A little paintwork on the wall

evening's fishing. Sea otters danced in the falling light, circling pools of water for the moonrise.

Monday morning and we pushed miles for a sunrise start. Soft to no wind, we were on the motor to leave Filomena Island and run back out into Canal Moraleda. Flat calm was something we had not seen in a long time.

With the newfound stillness, the water life was clear to see all about. The black silky torpedo shape of seals hunted their breakfast, Humboldt penguins bobbed about, and Albatross watched over from above.

Mankind in evidence, we had our best flag on the Jack pole for a nine-thirty flyby of a smart grey Chilean gunboat charging south down the canal. God and Empire, hoot the horn and salute the officers.

Close to the end of Patagonia, it was time for a last adventure. We turned a little east into Valac Passage for Canal Refugio and a beautiful route between the narrows. Sun in the shade with 3,000-foot mountains on both sides that were so tall they all touched the clouds. There were streaks of white rock in the dark grey walls of stone, and waterfalls tumbled from the skyline. We edged up this mighty ravine of nature.

Brakes on! Approaching Largo Island, Puerto Santo, there was a signpost in the rock at 43.57.80S 73.07.00W.

Passing yachts had climbed the steep bare slab wall to paint their names as calling cards. Painters are us, we dispatched the dinghy to join the list. "*Hollinsclough GBR 2009*" was written in red paint set in the middle of a history of travel — like you do.

Paint brushes cleaned, Bahia Mala at the north end of the canal beckoned. This location was direct from the set of the "*Sound of Music*" — epic Patagonian country. A sneak look at the Pacific swell coming in from the Gulf of Corcovado, and then we looked for the entry starboard side. Rock wall and shadow, we closed on the GPS into the darkness of the cliff walls. A zigzag entry sliced into the shadow, the gap only appeared when we could read "Made in China" on the exhaust pipe. It was a committed and close entry. Swell crashed into the rocks, and what passed through gave no option of turn.

Bahia Anihue – 43.52.30S 73.02.40W

We may have been getting brave, but it was a big sigh of relief to slip inside and fall behind Isla O'Brien into Bahia Anihue. In contrast, it was entirely sheltered inside. Tree jungle rose from the sea on all sides and touched the clouds. It was another breathtaking anchorage with common dolphin all over the bay. They chased the dinghy as we set a bow and stern line with the anchor for triangulation.

There was black volcanic sand for low water and a beautiful house on the shore. We were made most welcome by Francisco "Pancho", his wife Antonia, and their daughter Olivia Gomez Sepulveda. Gifts of hot bread and fresh eggs followed. Wow!

A few days of strong north winds parked us up to enjoy our newfound home. It was time for some maintenance. Engine oil changed, we re-greased the CV joint, and then scared ourselves to death. With all motivation, we had changed the diesel fuel filters, and we moved on to the high-pressure fuel filter and changed that too, but Polly Perkins did not want to restart. What scared us to death was that the starter motor then cut out. Overheated, an anxious hour later, the thermal cut-out came back in, and she fired up in a single turn. Good maintenance miles from help can be heart-stopping stress in a place this far away.

The magical Bahia Anihue, April 5th by map, chart and geography, we were leaving Patagonia. It seemed like a door closing, memories of a wild land of nature, remoteness, rainbows in the sunshine, and icy-cold water that we would never see in our lives again. From here, we were northbound for a left turn into the Pacific,

Anchored amongst the fishing fleet of Puerto Quellón

Easter Island, Fiji, Tahiti and northern Australia. The way home would be the Indian Ocean, the Suez Canal, and the Mediterranean.

Leaving Bahia Anihue, we steered by hand to exit the rock wall door back into Canal Refugio. There was trouble afoot. Engaging the autopilot, it had gone on the blink. For a shorthanded yacht, the autopilot is the third crewmember and cannot be done without. Nature helped, we turned into the open water of the Gulf of Corcovado and picked up a soft beam wind. It was in line with the swell setting from the west and took us easily to Puerto San Pedro on the southern tip of the large island of Chiloé.

A short run took us up the mainland-size island to the large town of Puerto Quellón. At 8,000 folk, this was our first big civilisation since Punta Arenas on the far side of Magellan — roads, cars and fishing boats everywhere. There was hardly room to anchor off the town amongst a myriad of workboats, fishermen's launches and small cargo ships. The lights of the city lay before us!

On the quayside, fishing dive boats were harvesting the most bizarre fur balls — to this day we aren't sure what they really were, but they were large mollusc shells the size of tennis balls. To collect them, six divers at a time walked the seabed, breathing air from long yellow hosepipes connected to a pump on the fishing boats 20 metres above. Collecting their goods, they proudly stacked them into vans on the quayside. We were treated to a tasting, fresh as you like. They had the consistency of raw mussels — yummy or yuk, not sure. Smile and wave!

Leaving the quayside, our prized booty was dinghy petrol, and an Esso station that was far more European sat at the head of the quay. A joy to fill up, we had motored our little Yamaha outboard engine RIB dinghy across Patagonia with no petrol fuel since Punta Arenas.

With a happy dinghy, it was up the hill to town. Black and yellow cabs buzzed around, and we found a bakery with marshmallows and a butcher's with mincemeat. A busy supermarket meant soft white sliced bread, eggs, cheese and ice cream. We found a large box of matches for the campfires and bags of toffees for toothache. The shelves were stacked with fresh sticky cake but no chocolate Easter eggs. Our bags were loaded down with booty, and the dinghy was full of petrol for our triumphant return to *Hollinsclough*.

It was time to strip the faulty autopilot; the problem lay with the control pack. The control pack was a computer in a box that converted data from the navigation GPS system to the electrical current driving the steering ram on the rudder. We changed the fuses, but nothing came out of it. We knew this was a key piece of kit, and we carried a spare. The spare was brand new — never been out of the box. It was an easy job to swap it over.

However, there was heartbreaking news on the autopilot, as the new core control unit had a fault. There was no clutch engagement to power the steering ram. Every fuse was checked, the cable run checked and re-checked, but after testing the new core pack at source, there just was no power on the clutch feed cables.

The city of Porto Montt was a few days away, so we would have to hand-steer the last legs and take stock there.

While the CV joint was sound as a pound, the thrust bearing fitted back in the ocean repair job at Puerto Madryn was rattling again. Let's not get paranoid, as it was a car bearing. In the safety of the bay, we slipped out the bearing and plate and took the bearing to Oscar at the car parts shop.

Oscar was delighted with the challenge and rose to the occasion, practising his English to help with an international repair. He found a monster wheel bearing from a truck and organised a local engineer who used his lathe to make more space in the mounting plate for the new bearing. Half the town seemed involved as

Oscar was delighted with the challenge.

CHAPTER FIVE: Pacific To Puerto Montt, Chile

A spring clean for the hull that scared the locals

the new work must have been ten times the strength of the part that came out of the boatyard when *Hollinsclough* was new. The *Armada* inspector got news and came to check. *"Chilean engineers, much grande."* He smiled and gave us a ship's MOT stamp.

The autopilot lay heavy on our mind. Via Blighty, Roy located a Raymarine electronics engineer in the city of Puerto Montt; we slept sound with that good news.

One last job was unexpectedly waiting. A volcano on the horizon was bright in evenings, its silhouette sharper for the light of a full moon. The full moon gave big tides, and by lunchtime the following day, the water drained from the whole bay. Fishing boats lay in the soft mud all around us.

Hollinsclough made 30 degrees on the horizontal wave dial; her keel sank into the mud as best as it could before she leaned over to sit on her side in the soft mud bottom. Careened by the ocean, we were dry on the floor.

The good news was that the hull was exposed from bow to stern. Taking such opportunity, the whole family launched into the mud. With gusto, we set about a hull scrub. We removed barnacles and weed from Patagonia. It was a wicked chance to have a spring clean and even change the propeller anode. As the sun began to set we had a very clean hull.

The locals stared in horror. The international yacht was on its side. A loud hailer from the fire truck on shore screamed out to *Hollinsclough* on the VHF. The *Armada* inspector on channel sixteen wanted to return and check us for damage.

The tide began to re-float us, and we had the new propeller anode in place — bolts as tight as you like by our own hand. We were dizzy with life in the mud, and footsteps aboard were slanted. Then we heard a suction sound the volume of a giant elephant burp! *Hollinsclough* left the mud and rolled back level.

We sailed the next morning on the sunrise of Thursday 9th April to bag a fabulous tide northbound up the Gulf of Corcovado, taking one-at-a-time turns on steering duty. The muddy job of hull cleaning had been useful, as we topped out at eleven knots in the current race. Flat, whirly water driven by the Pacific swell drove us north with force. Sixty miles of daytime steerage fuelled with fresh cake and toffees to chew in the sunshine — what more can a working crew ask for?

We could see the smoke plumes of an erupting volcano to the west as we turned in for the island of Mechuque. The church graveyard was marked with a green starboard tower on its corner. That's a seagoing burial for you. A town on stilts, the water was warm enough for a swim amongst the salmon farms and dolphins. Mechuque was a picturesque fishing village in stark contrast to the rock walls of Patagonia. Large white swans arrived for sunset; they had black necks and reminded us of the Commerson's dolphins. Fishermen had directed us to the *Armada* buoy, where the evening water was at its best. Moonlight rising, still calm, and then the town generator shut down — you could hear a pin drop.

A sunrise departure from the *Armada* buoy, we squeezed out between the salmon pens into the Canal Quicavi to run the island to port of us. Soon in the Golfo de Ancud, a flat water run with another good tide. The Straits of Queullin passage marked a gateway into the bay of Seno Reloncavi. A most unusual sight as rolling slopes of green grass hillside to our port side stood in full contrast of the mighty snow-capped Andes to our starboard.

Letters from home caught up with us at the yacht clubs.

Puerto Montt lay before us. The passenger liner *Norwegian Star* lay at anchor. We turned into a busy canal where fishing boats jockeyed for space on their run to the fish market. We came portside into Marina del Sur.

We hadn't seen the lights of a city like this since Rio de Janeiro. James Bond was not waiting; he was busy filming *Quantum of Solace* in the Chilean Arica Desert to the north. What did greet us was a full-blown European pontoon set, rope cleats and yachts all stacked shipshape. Our position was logged at Marina del Sur – 41.29.40S 72.58.90W.

The marina team of Marina del Sur were waiting for our ropes. 64-amp electricity, it was 220-volts, which was perfect for the toaster. We had mains fresh water at full pressure and a great pile of post waiting for us.

A cup of tea or open the mail? That's not an easy decision… get the mail opened. It was April, but what a joy to get the last of the Christmas cards.

We spent Easter Bank Holiday on the far side of the world with a city to explore and chocolate Easter eggs to find.

Chapter Six

Puerto Montt – Happy Easter

Chilean Military day, a foreign navy at its best.

"*The Legacy of Puerto Montt was the Spanish Language.*"

"*Ten miles up……. we could have been in the River Thames.*"

"*Weekend alone. We mulled over our route to be sure our sailing plan worked.*"

After more than a month afloat, we exchanged the untouched world of Patagonian nature for the lights of the city. People returned to our world, 200,000 people. City lights, a metropolis, a marina, engineers, school and shopping.

The city was Puerto Montt, and how quickly we made it our home. Our fixer, sailing guru and watch-keeper back home, Roy, had one box of spare parts already waiting for us at the marina office. Boat jobs it is then.

The autopilot electronics engineer was with us on the first day to look at the problem. It was the most modern kit he had ever seen: *"Never stocked one of those"* he stated, and our hearts sank. It would have to go to Raymarine in North America for service. Ouch, that came as a blow. Back in Blighty they would swap one from the shelf. Another possibility was to retrofit some older kit into our modern systems, but the different generations were struggling to work together.

With many emails home and a few days of thought, there seemed no option but to dig in at Puerto Montt and await autopilot parts from Blighty. To be sure, we chose to order a full set of autopilot parts from the core pack to the ram and head controls. Availability even in Blighty was a struggle; Andrew from Cactus Electronics in England sent some items and sourced other parts of the kit via a North American agent. It was to be sent direct to us. Import into Chile is all via the customs in the capital city of Santiago. The Chileans are not keen on imports, as they like you to purchase in the country. It takes weeks of customs exchange, so we buckled down to the idea of a longer stay and focused on what to best do with our time.

One legacy of Puerto Montt that would serve us proud was a vision of the Spanish language for the girls. With help of the marina staff, we sourced a language teacher called Claudio to formally deliver daily Spanish lessons. The girls were delighted with the prospect of hard work and a teacher.

As Claudio set about the language work, we took a look at some of the serious boat jobs. We hadn't been in a marina with such a wide spread of engineering availability in 10,000 sea miles — since Douglas on the Isle of Man.

Added to that, the marina manager, Rodriguez Rojas, spoke perfect English and was available for detailed technical translations.

The Pacific leg ahead would be thin on such opportunity, so work began.

First up was a shiny new red set of top-notch Trojan batteries. The battery box was re-worked to get the extra 50 millimetres needed to get them in. That facilitated a move for the fuel pre-filters to be raised on the rear of the battery box. It meant better access at sea for those drain moments of dirty fuel. The solar feed regulator was re-set and a new Cristec charger from Roy's parts box was put in place to see our electrics spot on.

A local firm had remanufactured soft rubbers for our engine mounts to make a whisper-quiet bed for Polly the blue Perkins. Two of the four old engine mounts had turned very hard — surely a secondary cause of all that vibration from the thrust-bearing failure. Chicken and egg? We would never be sure. Engine mount or bearing? One must have caused the other.

With new engine mounts, we swapped our home-engineered thrust bearing unit and CV for all new kit that had also arrived in the parts box. As problems had occurred, Roy had sourced parts in the UK, stored them, and together we had focused on a delivery to Puerto Montt. The parts box was a splendid opportunity to replace homemade solutions to items we had fixed on the hoof.

We stored all the old bits as spare. We were feeling very chuffed with the level of spare parts aboard.

Engine and generator oils changed, the diesel fuel filters all got changed on their new homes high on the battery box. We purchased twelve more pre-filters as spares from a local truck supplier and stacked them up in the hold. We had 40 litres of fresh oil on board, and new ATF oil in the gearbox. The engine was repainted and cabling re-wrapped with smart wide plastic ties.

A local rubber pipe shop helped with a variety of new hoses for the hot water tank. We found vent pipes for the freshwater tank and the anti-siphon tubing got changed. A new pressure escape valve on the hot water tank from a house plumber was also installed.

The VHF radio unit was re-wired to a new power supply. The mast cables were made tidier, and the joint box was replaced with new connectors at the base of the mast to be sure. We even got a new screen for the 12-volt computer from the local TV shop.

Lifting the lid for a porcelain polish was not enough, and we set about the sewage and waste water systems with a vengeance. The grey holding tanks were opened and cleared, toilet pipes stripped and cleared clean, toilet pumps cleaned, and all bilges washed through. One of the mid-generator bilge drains had long been blocked. It got unblocked, and all the pumps were stripped and cleared. Lovely job!

A good scrub of the yacht's bottom in the water supported our work in Puerto Quellón and another new propeller anode was fitted to be sure; we placed the slightly used one back in the spares pack.

Rodriguez found us a sail maker with a giant sewing machine. The cockpit sun canopy was re-worked and fully repaired from its sunshine baking at the Equator. A particularly special achievement was a new outer rain sheet fabricated to our design with water catchment in mind for the Pacific islands.

CHAPTER SIX: Puerto Montt – Happy Easter

Since the cutter sail had broken, its spigot back at Mar del Plata, the hydraulic motor that rolled the sail in and out had a small weep of oil. A local hydraulic engineer specialising in truck lifts took great joy in servicing the tiny hydraulic furling motor. "Never seen anything so small," he told us. They went on to change all the hydraulic fluids with posh Tellus 40 Shell oil. The bow thruster hydraulics even got a new primary filter.

The anchor was unchained and its stock straightened. A bracing bar welded onto the stock gave it a macho look fit for a ship. The splice joining the 50 metres of new chain from Uruguay to the old chain had always worried us, so we had the chain welded and even applied a new coat of paint to our trusty hook.

Yes, it was a mighty good session, and no stone was left unturned. All we awaited was the autopilot kit to finish the job. Customs was frighteningly difficult here, and UPS was taking an age to transfer goods from Santiago. There were a few weeks ahead for sure.

For Puerto Montt home life, Spanish teacher Claudio arrived every day, fitting in his Spanish lessons between our existing syllabus of Maths, English and Geography.

Life was not all school; fun on the water was a daily routine for the girls. They loved to dash about the enclosed waters of the channel in the motor dinghy and burn up petrol in the Yamaha motor. The marina became their gasoline alley, and they made friends with many of the local fishermen.

Morgause was the dinghy motor expert. With the sunshine strong and the water flat, she handed the control handle to Caitland. In the swap-over, the throttle went full open. Speed and turn together, Morgause was spat out of the dinghy by her sister. Never failing her position, she held on to the side line as the dinghy settled. Holding on was her downfall, a broken wrist the result. Time for the hospital then!

Arm in a pot and back to school. Cheering her up, the next history lesson was an outing to the newly opened

A Chilean blessing

Jesuit College of San Javier chums come for a boat visit.

city museum. Woolly mammoths to the present day, all the world but with local focus on earthquake devastation in the city around us. Following the success of the first visit, the girls regularly returned to the museum for their Spanish lessons with Claudio and became local residents of the library reading room.

English children in town was local news, and school invites followed. The girls' school of Immaculate Conception was first where heads of the local Girl guide troops were found and the head Sister blessed a new national flag for us to fly on the mast of Hollinsclough.

The boys' school, the Jesuit College of San Javier was the next to say hello. The senior boys visited *Hollinsclough* for a morning of lessons on navigation, ropes, rigging, stores and electrical systems.

Father Herman would then often invite us all for school tea and during one evening he showed us their very own cork tree.

Meanwhile, time to meet the Guia Scout Girl Guides. Wow, the blue mafia network was very strong here. We were just in time for the 100-year foundation celebrations of the Chilean Guias. Tracey was awarded the very first white birthday neckerchief in celebration of the 100 years of Chilean Girl Guiding. Morgause was sworn in as a formal Chilean Guia and awarded the country's colours.

It was the Girl Guides that signposted the next stage in the boat refurbishment. We were asked to host an afternoon tea party for the local *Brown Owls* (Leaders). Kitchen spotless, we even polished the brass doorplate trims in a hoover up and wash down

The Jesuit cork tree

CHAPTER SIX: Puerto Montt – Happy Easter

that saw every piece of bedding in a spin through the washing machine.

Another busy week followed. The Brown Owls escalated introductions as we met more folk from the church, the schools and the community. Everyone wanted to practise their English. Claudio was so successful teaching the girls Spanish that we realised we could give back some English teaching to the community. We helped with English classes at San Javier, the Jesuit School, from reception class, where we taught them to sing, "Head, shoulders, knees and toes", to senior boys' grammar in the sixth form. It was a whole lot of fun to put something back into the community that had taken us in as locals.

Girl Guides of Puerto Montt

Visiting new chums, Morgause acquired a local green team football jumper. After some weeks of away games, we had a full-blown, mean green Puerto Montt home match on the first Sunday of June. The stadium was just above the marina. We approached the turnstyles, *"Anglish, they are the Anglish, they are Anglish hooligans, let them in,"* shouted the policeman.

It was a three-thirty kick off, and the cheering was loud, drums banging and flares blazing.

For half time, the food sponsors sausage had taken human form and was in goal. Kick the sausage! Half time saw youngsters on the pitch for a penalty shootout. Morgause missed by a whisker. No goal! Hitting the sausage bang centre in the tummy, Morgause couldn't stop laughing.

Shipshape for the Brown Owls

The "Anglish" hooligans

A war-strewn terrace

The game finished two nil to our team, and but for the 10,000 miles, it could have been any Saturday afternoon in England.

With almost daily customs trips to the Aduana customs office, we were getting very close to the release of our autopilot kit. It was in transit from Santiago, and within a few days, it would be in the hands of the city customs officers.

Puerto Montt sat in the shadow of the Andes, and it seemed with a few more days in hand, it would be a great loss not to have a closer look at the mountains.

For a holiday in the hills, those very organised marina staff got us a silver Toyota hire car. It was inland for 7,500 feet of Volcano Osorno. It was a magic trip passing Puerto Varas to circle Lake Llanquihue for life in a log cabin at Bellavista. A wood burner stove meant sticky cake for tea. Our new A-shaped A-frame home sat on stilts above the world, safe from earth tremors and red hot lava!

The local farm worked lama and ostrich in lush green meadows rich in volcano ash.

The lake district of glacier heads, tumbling waterfalls and white rivers beckoned. The mighty Petrohué River swerved about twisty dirt track roads beyond the tourist village of Ensenada.

Our new A-frame holiday home

Steam traction engine relics were on display at many of the historic farm settlements of German immigrants. At Volcano Osorno, a towering masterpiece of Jurassic lava, we drove through the mist and above the tree lines to find a chairlift for an adventure to the top.

Upwards to touch the snow, the seats of the sky ride were made of last season's skies.

CHAPTER SIX: Puerto Montt – Happy Easter

It was becoming autumn in the southern hemisphere, and we were treated to a full-colour image in monochrome, black lava and white snow side by side.

The homebound option from the top of the volcano was downhill speed — a four-leg, 1,500-metre wire ride. A seat harness and carabineers formed a pull-down homemade brake. We scared the ranger instructors half to death. "Brakes on'… what are those words in Spanish? It was tremendous fun to whoosh down lines so safely attached to the ground. Crash mats and snatch ropes fit for the biggest anchor of any Patagonian storm waited at the bottom.

Some drainage by the roadside

Back to lake level, our holiday was over, and it was back to sea level in Puerto Montt rejuvenated and refreshed with our mountain adventure. It was brakes on for party time.

Late April signposted Caitland's birthday — twelve years old. That really is scary! The morning Spanish lesson with Claudio must have taken ten hours, but alas, the prize was a mega chocolate cake with twelve sparkling candles for a very English lunch with chums from half the city. The Marina del Sur pontoon team were delighted to conduct an array of visitors about the pontoons and showcase the marina to all.

The downside of all this civilisation and people was a shed load of cold. Thankfully not the flu ripping across the world, just a good dose of coughs and sneezes. The boat was well stocked with antibiotics and Lemsip, but it was time for the hospital. Not to replace the antibiotics, but to remove the plaster cast from Morgause's wrist. We visited Dr Marin on the fourth floor of the city clinic. He loaded us down with samples, smiled with glee, English, see. "Glaxo send us so many samples from Europe, take some!" We had pills for half the Pacific when we left his surgery.

Morgause's wrist clear of plaster, revitalised from a holiday in the mountains, the call finally came from local customs that our box of imported autopilot parts had cleared its paperwork.

We collected our Raymarine box at midday Friday 5th June. There was still a long morning of stamps and signatures, but we were victorious; there had been five office exchanges from Lan Chile to the port customs warehouse. Our goodies sat in a container with a padlock in the holding yard. To our horror, the box had been opened many times. Various customs tape and plastic re-wraps adorned the corners. Home to the marina office, and to our delight, all looked well inside.

Like Christmas with new toys, we fitted the core pack computer, new electric compass, new feedback to the rudder and placed a new steering ram on the rudder stocks. With the new parts fitted, we had spares for the lot. The cabling was very tidy, and the core pack ran two communication lines. We dedicated line one to the head steering computer and, more conventionally, put line two into the instrument ring. This disconnected the power return. If a separate instrument failed or spiked on the main ring, we should still have an autopilot. That's a spare and a protected back-up to be sure.

With the entire kit of autopilot new, there was about 1,000 pages of sea trial and commission instructions. Fingers crossed please. We would run some sea legs and turns to set steerage to balance the rudder and match the sea state.

Zarpe transit papers done with the *Armada*, it took longer to pay than prepare the documents. No cash was allowed and no cards were taken. We had to visit the bank to transfer funds. How hard is that in a foreign language? The girls put their newfound Spanish to hearty use. For 90-day passport stamps, it took two visits to the International Police; "International" was on the badge, but they didn't speak a word of English. Then with our passport extension sorted, we needed another 90 days on the Aduana customs documents for the yacht. We had never been in one place long enough for such excitement.

For one last bottom scrub of the boat, we had two wetsuits on, one above the other, as summer was ending. We turned blue in the cooling autumn water. The hull had a dark silky grease coating from the neighbouring fish processing factory. The antifouling equipment and the anodes remained in very good condition. We used an hour's air on the dive tanks, and the air had to be refilled. Air refills in South America are limited to 200 bar. Our tanks being certified for 300 bar euro pressures, they are left a third empty when re-filled in South America. A great result was finding the shop called "Ecosub", who sold us a spare twelve-litre second-hand tank for 50 dollars. Bargains are Chile — bring those Pacific reefs on.

We were sad in our hearts to face the reality of leaving Puerto Montt. A cottage pie lunch of community celebration with the Jesuit priests at San Javier school ended with bread pudding and was a fitting last supper.

There had been much work on Girl Guide badges with the blue mafia of Puerto Montt. The excitement was building an outdoor shelter for the survival badge. Robinson Crusoe Island was ahead — surely the place to build an outdoor shelter.

Bon Voyage! On Thursday June 9th, we left the lights of the city of Puerto Montt.

CHAPTER SIX: Puerto Montt – Happy Easter

It was such a shock to be back on the water, but we made a very easy first-day motor run past the pretty town of Calbuco. We were delighted to report a pelican fly by, then down to the entrance of Canal Chacao — the gateway to the Pacific.

Vibration, oils and water all good. Autopilot responding and turning to the sea state beautifully. The flaring propeller was a little sluggish — maybe too long parked up.

First, an evening on the chain at Abtao. We sheltered in a horseshoe of hillside lying behind the salmon pens by some local fishing boats. Friesian cows were herded for their evening milk on the shore. Anchorage bueno! The water was so much cleaner than the city that the new paintwork on the anchor was visible in ten metres of water. *Hollinsclough* sat nose into wind, natural as nature intended, windows drawn in. A day from the city and how quiet we were, shipshape and fully repaired. The plan was a short run north to the city of Valdivia and then a big left turn for Easter Island.

We co-ordinated leaving Abtao for a fast tide run through fifteen miles of the Canal Chacao into the open Pacific. We maintained a shockingly quick twelve knots average ground speed for two hours. It was a rocket speed slingshot into the swell that was delightfully soft. Soft Pacific, that was a shock indeed — luck was turning to our side.

The wind forecast was good to push us on an easy run to Valdivia. Motor still on, 1600 revs for seven knots was a good safe shakedown. The prop was sluggish to engage, but it was OK. It could have some weed tangled in the blades, but it should have gone by now.

During a long, clear evening of giant full moon at sea, every inch north was a little warmer. At sunrise, we thawed out and celebrated the end of the forties latitudes. Yes, 39 south was on the dial as we turned into Puerto Corral estuary. Wind against us, the cutter sail helped the motor as we tacked into the head of the Valdivia River. Small open yellow fishing boats were all about us.

Ten miles up, we could have been in the Thames. The marina of CSV Yates in downtown had a posh clubhouse, showers and toilets but there wasn't a soul about. We took the best mooring on the block, right at the front. We tied to the pontoon rails and put a spider web of rope together, giving rise and fall with the tide.

Scull rowers arrived in the evening sunset, water still mirror calm. The Valdivia River was a splendid spot for a few days to wait and judge a perfect weather window out into the Pacific for that Robinson Crusoe Island Girl Guiding survival badge.

June winter was arriving in the southern hemisphere, and here in Valdivia, the Pacific cooled on the mighty sea cliffs of middle Chile, and it rained. It rained, and it rained some more.

The river city of Valdivia

Valdivia was a modern city and there was much to explore between the showers. The whole place was rebuilt after a monster Richter scale grade nine earthquake in 1960. By our standards, everything was new. After eight weeks in Puerto Montt, it was a welcome change to the windows of *Hollinsclough*.

We were eager to make Pacific water and escape this land-driven rain. By morning, the dive tanks were on and we were under the boat to look at that propeller — mirror finish and not a weed on it. The design of the propeller allowed the blades to flare out for motoring and fold back and slip through the water for sailing. Each of the three blades was on a bronze gear and packed with grease. Club Yates had a drying slip, so we would drag out in the morning tide and re-pack the propeller grease.

With *Hollinsclough* on the slip, a local boatyard engineer familiar with the flaring prop system arrived.

He took great care to match the blade gearing as he stripped the bronze parts. The grease was as full as new, and the gears hardly had a mark on them. Delighted and heartbroken, the problem must lie in the engine's gearbox. It was a quality Hurth unit, but there would be no spares for this sort of kit in South America.

The timing of the failure after sitting in Puerto Montt was unbearable.

Our experience of imported parts was firsthand; it took weeks to get imports cleared into Chile, and our window of Pacific seasonal weather was closing by the day. As much as we grabbed by the fingernails, Tahiti and Fiji were slipping out of our hands to the oncoming Western Pacific cyclone season.

On the slip for a look at the propeller

CHAPTER SIX: Puerto Montt – Happy Easter

Hollinsclough back into the water, the engineer was quick to get aboard for the next morning; our problem was not prop flare but slip in the gearbox. The unit was soon removed and in a workshop.

It looked good, as the problem lay with the slipping plates. New thrust bearings could be sourced locally, but the plates would have to be imported. Hurth ZX gearboxes were typically imported from North America but were made in Italy. We had become experts and clever import logisticians, but this was still no small task. The speed lay in the import paperwork. It was time to look to Australia. A Brisbane Hurth agent had the plates and could ship on a Pacific flight to Santiago. The reality of Australia as local confirmed that we were on the far side of the world.

With opening hours and time zone variations, it took a week to agree the part numbers, transfer a payment and confirm a carrier for customs. It would be two more weeks to clear Santiago Aduana customs our end and transit down to Valdivia. Internal courier delivery in Chile typically runs on the bus networks.

Re-grouping our thoughts, we chose to consider the worst, expect delays and dig in for life in Valdivia much as we had in Puerto Montt. School, Girl Guides and Spanish lessons… even another holiday maybe?

With help from marina captain Edwardo, a Saturday afternoon trip to find the local blue mafia Guia Scout Girl Guide group was organised. The patrol huts were beyond the bridge by the German school, neckerchiefs of purple and gold for fun with games, chanting cries of patrol and much badge talk. We gave them ladders and hand games, and they gave us friendship and an open door to the weekly pack meetings.

Velcro goes down at a pace, and Guia Scout Girl Guides sorted, it was time for school. We introduced ourselves to local lessons at Windsor School — logo of the lion. The head teacher, Miss Gloria, invited the girls to class for the last weeks of the winter term.

Three weeks of Spanish, Spanish, Spanish and a little maths and English. Hockey, flute and violin for extras made life so normal the girls believed they were back home at school in Repton, England. The head of the English department at Windsor, Miss

Back at school for an awful lot of English

Could have been in England

Cindy, took on the private Spanish tuition Claudio had started, and we had linguists in the making. Parents as like parents do, we soon got interlocked with helping out at school. We took class to deliver English as we had at Puerto Montt. Heroes of the native language, we taught nursery rhymes to infants, history to the prep, and life in Europe to the seniors.

So successful were our teaching exploits here that a gala dinner was organised for us by the teachers. With news of native English speakers on the jungle telegraph, and the next moment found us, family together, delivering some free lessons to the university students of Universidad San Sebastian. Our Blighty English was a rare continent apart from the typically American twang delivery of the North American taught locals.

Amidst all this excitement, the pain of gearbox failure was subdued with the strength of a return to city life.

In the lights of the city, we were fast becoming Valdivia experts. Famous for its beer production after German colonisation, the yacht club was a short walk into town past a splendid old round tower fort. The tourist river front ending with a fish market bustled with bargains, and sea lions were fattened on the fisherman's leftovers. A modern casino hotel, skyscrapers, then the main river bridge past a Terpel petrol station. *"Tin of Terpel for the outboard motor please and a pint of beer to finish the dinner!"*

We visited the Anwandter House Museum for a history of the town's colonisation. Lord Cochrane display, remember Aubrey,

Schoolteachers arrive for pre-dinner drinks

CHAPTER SIX: Puerto Montt – Happy Easter

Russell Crow, Master and Commander? he was based around Cochrane on the far side of the world. Hold Fast! Days out following Lord Cochrane for a number 20 bus trip back along the north shore of the river to Niebla. Spanish iron and long nines lined up for a defence of the estuary entrance in a splendid headland fort we had seen on the sail in. The gun emplacements were carved out of the rock, making modern wide walls of artillery defence.

World War Two movies were before our eyes.

We had memories of a similar place in Brazil called Fort Orange. The Chilean fort was not Dutch though, as this was Spanish territory. They held until Lord Cochrane took the place in the 1800s. Hold fast indeed. There was a splendid museum of exhibits from Indians to Spanish galleons with maps of a changing world from Magellan to Drake and pirate Captain Jack.

Amongst the tourist excitement came Armed Forces Day, and the army navy and air force marched the streets of Valdivia. The navy was hardly distinguishable from Blighty, but the army was straight out of a World War Two movie of storm troopers.

Like Puerto Montt, it was time to visit the football. Hooters loud, Valdivia was graded division three. They wore red and white colours for a one-nil win with a nail-biting finish.

Across the way was a Blockbuster Video store, McDonald's and a modern mall. There was also a supermarket called "Unico" and another called "Bigger". We found an Esso petrol station for 50 litres of diesel to bring home on the bus with some ATF oil and packing grease. Are you allowed diesel on a bus in Europe? It was a small world when we found that the owner of the Esso petrol station owned the yacht next door to us at the marina.

We made regular trips for tea at the Entrelago chocolate shop café, had beer and steak at the German Kofman brewery and fried pasties, broth soup and corn pie by the market for dinner proper Chilean.

Each day when the river tourist boat sailed by the marina, it was deck duty. We had the hooters as loud as the football and gave waves to tear our arms out as *Hollinsclough* became part of the tourist river trips.

Life in the capital city of Santiago

Chilean medical services were very cheap. We had fillings checked and teeth polished at the dentist, a trip to the doctor for more antibiotics and even a session at the Case Military hospital for a big toe fix. For English visitors they turned out the colonel, like you do! Medicine is the cheapest we have ever seen in the world; we paid a two-dollar hospital bill for a bandage on the damaged big toe — doctor and nurse all in.

Between the excitement of food and fillings, school chums said it was time to go inland and sleep on a bus. Five hundred miles overnight through the Chilean vineyards for a week of capital city life in Santiago. Club de Union and English life abroad with five course lunches below ceilings of Tudor carvings above Italian marble floors was a long way from sailing oceans.

We had ever more Guia Scouts to meet, International Director Pia Valdivia and Guia Secretary of Chile Bestabe Irigoyen. It was perfect timing to meet the Queens Girl Guides from England on their Chile Gold project — Charlotte, Andrea, Lindsey, Sam, Rachel, Helen and Kirsty.

We took a view of our future travel at the Rapa Nui exhibition in the palace and a walk through time at the pre-Columbian art gallery to see the work of the Aztecs. Santiago was a city of smog with fantastic views of the snow-capped Andes Mountains all around. The metro underground was like London, and we enjoyed a hillside view above the Covent Garden of Bellavista. A funicular to the hilltop statue Immaculate Conception for prayers on Sunday was an interesting experience.

But for the reality of geography, we could have been in Lisbon or Rio. From the lights of the capital city, it was bus home to Valdivia for the first days of August.

The gearbox plates were finally in our hands.

In sterile conditions, the engineer replaced the plates. They were technically like a motorbike clutch, but in tandem with two of them set back to back. We remounted the box, new ATF oil and ran it up under pressure on the ropes — all was well. Mechanical luck had not shone well over *Hollinsclough*, as the sequence of events from thrust bearing to CV joint had clearly stressed the engine mounts.

CHAPTER SIX: Puerto Montt – Happy Easter

How much effect that had on the gearbox and the rate of wear was impossible to judge, but there was a whole lot of coincidence. It was just bad luck. There would be one more possible piece in that jigsaw, but that lay ahead.

We were ready to sail. We were ready to continue our quest to answer if the world is round, but the season to transit the Pacific was lost. Polynesian Cyclones on the Australian side of the Pacific are to be avoided at all cost in late season, and that was a direct clash with the arrival times of our Pacific transit for this year.

Boys and toys, but we do have the Loctite.

Our world tour needed a review to its schedule, but the weather seasons were not our only problem.

The general plan of our quest was to sail the world on the old wind routes. Our trip motto *"Is the world round?"* should have seen us return home via mid-Pacific, Tores Straights north of Australia for the Indian Ocean and home via the Suez Canal. Somalia and pirates of the Red Sea had closed the Suez route for all but the very brave. It was not impossible to reroute south, but the Madagascan Channel and the Cape of Good Hope was no place for a family yacht. The pirates had closed our door home.

Parking up in Australia for a year to see if Somalia settled down was appealing, but Caitland was set to start senior school the following year, and the term commute flights back to Derbyshire would be a huge distance. If the Somalia politics didn't settle, we would be parked up for some time and committed to a lot of long-haul air mileage.

It was here in Valdivia that we came up with a variation to our master plan. Sat in the Pacific, we would use the last of the season's winds to sail out to Robinson Crusoe Island — we had to bag a Pacific island. As the end of season winds turned, we would come about and return down the Pacific Patagonian coast, where we had gained an addiction for the beauty of the ice. We'd then take the Beagle Canal for Cape Horn, trade winds for Christmas in the Falklands, visit the jewel of South Georgia for some Antarctic adventure and carry those trades to South Africa. We would be in Cape Town before September and be on the BA Heathrow flights for school. *"Is the world round?"* could continue with a mid-Atlantic run via Ascension

It's a family around the world as close as sailors.

to the Panama Canal and take up the Pacific in a year's time from the north. That would give us as much as three years to see Somalia settled.

It all looked sound to us; it fitted the seasonal wind runs and brought sense to the dilemma of all our mechanical misfortune.

The girls had a weekend camp with the Guia Scouts of Santa Innes; the old bus hired for the trip was air conditioned with holes in the floor. We joined other parents and helped push start it to overcome the flat battery.

We had the weekend alone to mull the route and be sure the sailing plan worked.

Time to think with the grey lady

CHAPTER SIX: Puerto Montt – Happy Easter

We took a dinghy trip up past the submarine to focus our thoughts. As the grey lady sat there on the town quay, we wondered if that crew could answer our question: "Is the world round?"

Family back together, minds set, boat as well as it had ever been, we said goodbye to Valdivia and sailed for our first Pacific island. Our minds were focused; we would build a shelter on Robinson Crusoe Island for the Girl Guide survival badge, come about and sail for Cape Horn.

Chapter Seven

Robinson Crusoe Island – South Pacific

Robinson Crusoe statue.

"*Our minds were alive with adventures of the south seas…*"

"*By the third day we were feverish to move.*"

"*Morgause defended England's tradition of sailors…*"

From landlocked repairs and city life, *Hollinsclough* returned to the Pacific Ocean with a mission to bag a Girl Guide badge. We tore the Velcro fasteners from the river city of Valdivia and swapped marina life for sailing in the swell of the mighty Pacific Ocean. Our quest remained focused on finding out if the world was round.

We were deep-ocean sailing in the waters of the British privateers, buccaneers and pirates, in the footsteps of Bartholomew Sharp, William Dampier and George Anson. The territory of the Manila galleon, square master ships of gold and treasure on the Spanish run home from the east.

Sailing amongst the history books, *Hollinsclough* took all the square masting she could, jib sail

Landfall was never more famous than this.

poled tight as a drum, cutter opposite to the jib, feeding the wind, and plenty of main as wide as you like. Our stomachs settled to the swell by every mile covered, and by three breakfasts, we were acclimatised as a distant island climbed out of the horizon before us and a new anchorage loomed.

Juan Fernandez Island, South Pacific – 33.38.40S 78.49.60W

We arrived in Cumberland Bay for lobster season October 1st. Friendly fisherman in long open boats guided us to the softest water and watched our anchor down to safety. There is no better feeling of security to have the best local knowledge show you where to place the hook.

A tremendous horizon of mighty jagged volcanic mountains was all around us. Their two-million-year-old fingers stretched into the clouds above a dense carpet of deep green trees. We were surrounded by the sweet smell of sandalwood, enormous rhubarb-like ferns, and firecrown hummingbirds flew amongst the rainbow of petals.

Some reality of the fragility of this place was signposted by a tsunami evacuation sign. It pointed to a track, a straight line of Roman accuracy, cut bare of vegetation for height above the village in a storm.

Volcanoes in the mist from Cumberland Bay

As we stood on the landing platform, we could see thousands of fish enjoying shade under the quay in crystal-clear water. The singular town is called San Juan Bautista, Espaniola for St John the Baptist, the same name as the church back home in our Derbyshire village — there's fate for you.

We arrived a day after the arrival of the monthly cargo ship, so everyone busy ferrying supplies ashore by big RIB from the cargo ship. There were wheelbarrows all about the place to carry the booty home. Gold and treasure was ditched for fresh vegetables, meat and potatoes.

This was the real live home of Alexander Selkirk, marooned for four years and four months, dramatised by Daniel Defoe and embedded in the minds of every English schoolchild in the name of Robinson Crusoe. Three hundred years after his recue, Juan Fernandez Island was renamed "Robinson Crusoe" in 1960. His statue stands tall in the village square and has enough inspiration to fill the heart with the wonder of this remote paradise, rarely visited and with no connection to the tourist route.

An evacuation route that would be used that year

Our key task was to claim a Girl Guide survival badge. The

CHAPTER SEVEN: Robinson Crusoe Island – South Pacific

badge requirement was to build a survival shelter in a remote location — where else in the world? A giant tree provided an arch of its branches to support a roof of undergrowth and bracken that made us snug for a castaway experience. A Girl Guide flag for a wall, and we had a survival shelter of Crusoe extravagance befitting the survival badge.

Communications and fishing boats for San Juan Bautista

An *Armada* Toyota 4x4 with raised wheels was the only vehicle on the island. It was perfect for the narrow concrete road and acted as a support vehicle for port control, police, navy and mayor. We met the school headmistress, and as we had done throughout our trip, we took English class in the school of 150 children, class size of ten or so. They did a special English presentation for the girls, and then it was best togs for breakfast with the head. Sunday, we met the vicar, who is also the postmaster and himself a keen Guia Scout. We exchanged Guide badges and enjoyed a lobster pasty lunch. Lobster and eggs for breakfast dinner and tea. Lobster, lobster and more lobster. They were crawling out of the water onto our dinner plates.

We began an exploration of this fabulous island and walked above the village to the old Spanish fort. Some ancient caves were found, a modern home to the patriot prisoners of Chilean independence.

We enjoyed picnics in paradise with soft sunshine — cool and without bugs, but bright enough for suntans.

The true mountaintop lookout of Robinson Crusoe was a day's walk twisting through the forest with a plaque of commemoration at the top. Our hearts beat with the height of the climb, and our minds were alive with adventures of the South Seas.

A day back at sea level, island dive instructor Pedro Niada kitted

Hollinsclough lies in Cumberland Bay below

There were no footpaths.

us out with steamy seven-mill wet suits to snorkel with the fur seals. We joined forces with Bart Veldink on his sail yacht *Tranquila* for a navigation of the island and the north cove. The sea lions were almost extinct, killed for their skins by the million in days gone by but recovering in what is now a World Heritage Park. It was a privilege on a grand scale to swim with these wild creatures in their remote island home. These friendly seals ill prepared us to meet the aggressive fur seals of the southern ice, whom in enormous contrast you defended yourself from with a large stick.

Like the seals we would meet in the future, the weather here became angry in a moment; an evening's anchor in a wind turn became a 24-hour watch. With all the local help in the world, the bottom was hard and rocky and would spit the anchor free with little warning.

On a particularly blustery day when the wind moved across the tide, we drew and dropped the chain three times. We cooked one of the electric motors that drove up the hydraulic pressure to the anchor windlass winch. We had two motors on board, so we swapped the spare, but it is never the same when you haven't got one waiting in the locker.

Robinson Crusoe Island, Girl Guide badge achieved, and a full recognition of the fragility of the environment, it was no place to take a Velcro holding.

It was a sad day to draw our anchor from Cumberland Bay and say goodbye mid-October and we were away. Bart was sailing *Tranquila* north towards the Panama, and we sailed East for Chilean mainland and the port of Concepcion to follow the vision of our new master plan with Christmas in the Falklands.

Two Chilean *Armada* warships were manoeuvring off the island, both mighty type 23 frigates dancing grey shadows in the sunshine — what a sight!. We called VHF 16 to say hello. FF07, the *Admiral Lynch*, returned our call, wished good wind and following seas: "Any problem just call us."

Acclimatisation from the anchorage to the sea was much easier and breakfast was a hearty feast on the water. Life asleep in a rough anchorage leaves little change to the motion of life under sail. The Pacific swell around us was big but calm in its size, and we welcomed our return to travelling on the waves.

CHAPTER SEVEN: Robinson Crusoe Island – South Pacific

We had two starlit evenings of spectacular ocean sunset, the like of which is never matched from land, not an aeroplane line in the sky. The season's wind turn had been kind, and *Hollinsclough* was making fast downwind sailing miles over ground as quickly back as she had run out.

Closing the coast, we ran into a thick fog. Technology time, steerage on radar and plotter west of Quiriquina Island, visibility close to zero to find the outer orange buoy, and then spinnaker pole down and rigging away to motor into the Marina Manzano at 36.41.20S 73.06.15W.

Concepcion is not really a port, as the whole zone is a Chilean naval base Talcahuano. Updating our *Armada* zarpe inside the security of the red helmet Police Militare, we received dockside passes and navy security clearance.

This was life in a military base, but it was still sunrise to pelicans and sea lions. *Armada* Chief Petty Officer Carlos Ojeda organised a VIP tour of the historic 1814 iron gunboat RH Huascar. Arturo Prat captured the Huascar from Peru and later died on the iron deck, navy warrior and hero of Chile.

A special tour of the dockyard, coffee and ice cream in the naval café, photos of visiting warships, and a history lesson of honour followed. Former *HMS Sheffield*, now Chilean FF19 the *Admiral Williams* berthed on the concrete, bow like a razor, decks adorned with technology and missile. Beyond, in retirement, lay a Leander frigate chained to a destroyer, and beyond that lay the towering wooden masts of a square-rigged training ship *Esmeralda IV*.

The city of Concepcion was half an hour up the road. Sheila Gallardo from the yacht club and her husband Marcello guided us around. Modern Malls, stores, Riply, Johnsons & Paris, Lider and Sodimac lined wide streets. Fashion and technology mixed with a local food fair by the town hall and a mega-burger dinner at Mamuts. Sunset back in the bay, and there was no forgetting navy life as a Scorpion class submarine, the SS-22 *Carrera*, made port before the rise of stars. There was no doubt we were enjoying life in the navy.

Saturday was regatta day. Timing and parties are us!

Fifteen multi-class yachts were rigged for action amongst the warships. Quiriquina Island is the Chilean *Armada's* basic training school for ratings, Escuela de Grumetes, and the regatta was set for a lap of honour to finish on the naval school quay like a tour of the Thames for dinner at Greenwich.

With the start gun fired, an *Armada* RIB rushed us out to the *Armada* pilot boat.

From there we were given a VIP tour of the sailing fleet, modern Kevlar sails and spinnakers alongside classic wooden sloops — what a delight! Leaving the competition, we made for the finish. Navy blue, spick and span, a thousand sailors all

Sailing chums for the regatta

about us were proud to wear the three black ribbons of Nelson's battle honours around their collars.

Sailing travellers from England, 12,000 nautical miles away, we were given a grand welcome. Morgause bagged the job of the finishing siren and eyed the line. The base commander, Jaime invited us on a tour of the base. Escuela de Grumetes "EG" gold pins on our lapels, it was a great honour and a fantastic tour. The next invitation was to the celebration hamburger lunch for an enormous stuffing of meat to celebrate the sailing.

Two hundred and fifty tons of grey *Armada* ferry took us back to the mainland quay. We enjoyed a tour of the ship's bridge, and Morgause took the wheel for a bit of ship steerage. Two hundred and fifty tons and all, what is she like? Does that count towards her junior inshore RYA powerboat ticket?

Nelson and Trafalgar history formed the platform of the Chilean *Armada* and Talcahuano Naval Base. The *Armada* alongside the yacht club provided a military adventure of grand proportions.

No matter how bad the storm of an arrival, it's always the leaving that is harder. We set southbound and again those seasonal wind turns favoured us. Big pelicans escorted us out en route to a fishcake dinner; it was no match for the *Armada* School regatta lunch we would remember forever.

Three days down the wind followed — easiest sailing in the business. The jib was on the pole, main as far out as the ropes would stretch, and the cutter was backwards as a giant duct for the jib. It was a lumpy bumpy run on the turning water, all sail and no engine, so we were saving electricity to power the au-

Hollinsclough in our wake for a regatta with the Armada

topilot, which was working hard for a living, steering fast to adjust to the sea state as waves bounced at the bow from port and starboard on a fast run. The inland waters of the Patagonian canals beckoned us back to our anchorage at Abtao.

We had local knowledge and tide times to perfection as we turned into Canal Chacao. It was all speed ahead with five knots of tide push, no fog but darkness and no moon for a blind arrival, needing all our technology to arrive in the narrow channel on the radar. We used the previous plotter coordinates to drop the anchor in the beautiful and protected bay of Abtao. Our very first return anchorage! Silence, stillness and without moon in the cloud, a darkness darker than a tin of matt-black blackboard paint greeted us with an unnatural stillness after the Pacific swell. Our position was plotted as 41.48.14S 73.21.64W

Two hundred and fifty tons, and Morgause ready to take the wheel.

We had a morning surprise the next day when the engine wouldn't start. Where did that one come from? This would be the last legacy of the thrust-bearing failure that began on the Atlantic side of South America back in Puerto Madryn.

No diesel on the injectors meant a day of fuel bleeding first, a dirty, stressful job that took a toll on the starter motor. Polly Perkins was having none of it, so it was time for a call to the Chilean *Armada*. A clipper, *Caucau*, arrived by sunrise the next morning, and the engineering officer diagnosed the fault as the high-pressure pump. They don't last forever, but they do last a long time. In our mind, this was the last victim of the vibration. Propeller shaft thrust bearing or failing engine mounts? Chicken and egg… we will never know, but it had cost us dear in time and distance.

The pump was withdrawn from the engine and despatched with all speed to engineers in Puerto Montt for repair. This left us with a couple of lazy days of sunshine in the flat water of the crystal-clear bay. It was only a day's sail to Puerto Montt, but the run turned direction about the many small islands and narrow channels and needed a good wind to sail.

By the third day, we were feverish to move; we could hardly bear the tranquillity. Some sailing wind came in our direction, so it was anchor up and away. It was a zigzag run between the islands, and close on Puerto Montt, we found some towing help from a local landing craft to see us head into the wind and come around Tenglo Island. We made Marina del Sur for the friendly engineers in the early evening. We soon had shore power and lines to a pontoon with all the water we could use. We then made a return to the mall, which meant shops of the world and white sliced bread.

The high-pressure diesel fuel pump is like a little engine all of its own, and the good news was that this was diesel technology used the world over. A LUCAS CAV unit, not even Perkins parts, was in South America and would arrive in a few days on the trusty bus network.

News was about that we were fresh from the *Armada* school at Concepcion. The Puerto Montt *Armada* Regatta was starting, and invitations followed. It was mainly big boats for the sailing race, but Marina del Sur took great pride in preparing a single-handed race sailboat for little Morgause. She became the hero of the marina with her fearless practice runs across the pontoon space. There was no broken arm today; she had mastered her sail-racing skills back in Puerto Madryn on the Atlantic coast. That location brought some irony to the beginning of our mechanical misfortunes. Her success would surely see an end to our bad luck.

The main regatta course was around widely spaced cans in the ocean of the main bay, legs as far as you could see. There were many single-handed Laser race boats for the big beefy macho sailors, but sadly, and not surprisingly, no young Optimist racers turned up. The *Armada* decided to place Morgause in her Optimist alongside the Lasers in the main bay. She was like a toy bobbing about in a bath tub. Her sail number was proudly re-written from 7777 CH (Chile) to 7777 GB using some black insulating tape. It was time for the top brass to take a look.

The four bar *Armada* Captain Pablo Muller Contreras invited the English sailing parents of this intrepid child racer aboard his 500-ton ship *PSG Micalvi*.

We were given five-star treatment to watch the event from the luxury of his control bridge. It was splendid fun, and tremendous lunch finished with Remy Martin, as the *Armada* do.

Meanwhile, Morgause defended England's tradition of sailors and, with some astonishment to us all, she came inside the cut-off time of the fast Lasers to get her medal.

A GB sail racing number

CHAPTER SEVEN: Robinson Crusoe Island – South Pacific

Four big laps and a million miles on her Optimist log — bless. Adrenalin driven, and wet suit or not, she turned blue.

Every Navy *Armada* team was on standby for the top brass and had a French Dauphin helicopter for photos. "We will keep an eye on her," said a huge lieutenant commander. With red overhauls and parachute wings, he could have been John Wayne. He had a team of sailors in an orange RIB. Navy Seals, as they do best, took the opportunity to secure the international visitors' safety. Bar of chocolate in hand, not a problem, we would warm her up and feed her.

Captain Pablo Muller Contreras and his staff

Morgause Lomas, Optimist ocean racer extraordinaire — Puerto Montt gold medal winner. What a hero, and surely she had broken the run of mechanical failure bad luck.

It was a posh ceremony for those left standing after the lunch. *Armada* Captain Pablo Muller Contreras presented his international entrant with a specially bound history book of the Chilean *Armada* — what a prize! The first to sign the cover was the *Armada* Seal commander.

Like the *Armada*, life in the city was almost overwhelming with kindness and friendship, and we returned to the Guia Scout groups of Immaculate Conception and San Javier. Time was spent with the English students at San Sebastian University, then it was the end of term, and we celebrated the southern hemisphere year end with many chums; warmth, food, comfort and friendship was so strong we could have been local.

It was time for the city marathon. It was a fun race for youngsters and family, and we found ourselves standing as a full team on a start line. With the sound stage and warm-up dancing, we could have

Morgause gets her very own navy seal.

We will soon have you back to the party.

been at the London Marathon. Three kilometres was the length for the family run, with free entry and free red logo T-shirts — it had to be done! How surprising to find we were the only international entries. A hundred miles of energy later, we crossed the line British heroes. Orange juice and cereal bars were given to aid recovery, and we wore our red T-shirts with pride. Get those medals on the wall!

Heroes of the marina, the pontoon team had nothing short of a ticker-tape parade for the little girl's return — the children who raced sail yachts and ran marathons.

With word about the water world of Puerto Montt, the diesel engineers were back. The injectors were cleaned for good measure, oils changed and pipes and filters all fresh. The fuel pump was brand new, the timing was set with meticulous scrutiny, and Poly the blue Perkins roared back to life like she had never been away.

The fuel pump delay in Puerto Montt opened a new door, and Caitland made a solo flight to Blighty for her common entrance school test. That meant Chilean bus adventures and British Airways out of Buenos Aires — 25,000 miles. It was hard to be sure who was the greater hero, Morgause on the world sailing or Caitland on the world schooling. Caitland had an extra prize, as she returned with 25 kilograms of luggage loaded with mince pies, ginger nuts and secret Christmas presents. There was even something for Polly the blue Perkins engine. She had a hot off the press brand-new ultra-high-power 24-volt alternator for the engine. Well done our English fixer Roy back in Blighty. This would push power back into the mighty new Trojan battery pack at 60 amps at a go for short engine runs and put our battery management issues to bed. Between soft solar feed, the wind turbine and the big alternator, we would never have to save electricity again.

Christmas was looking good, and Father Herman of the San Javier Jesuit School had one last treat lined up for us. It was one of those moments to touch the soul, as he organised a visit to meet the Sisters of the Convent of the Carmelites — special prayers for the trip south. And touch the soul they did. We were in the presence of six Carmelite nuns; they had a tranquillity beyond the world, and their prayers for us inspiring to the heart. *Hollinsclough* had been blessed in St Katherine's marina

London by Father Paul Baggot of Holy Redeemer Church Clerkenwell. With the South American blessing of the Carmelite sisters, *Hollinsclough* was sound for a trip to the end of the world.

She was filled to the brim with fresh diesel on the commercial dock, before we sailed from Puerto Montt, returning to the epic vistas of Patagonian, snow-capped volcanoes, cloudy mornings warming into afternoon sunshine and all ice blue.

Puerto Montt fuel

Our first evening was easy on the *Carabineros* (police) mooring buoy at Isla Mechuque in the company of the wild swans. The red port marker edged the graveyard with an intent that it would be there forever. Locals smiled as they passed by on the shoreline, and we slept easy with the clarity of a plan in our minds.

Leaving Mechuque, there were many Humboldt penguins enjoying Pacific fishing in the stillness of the Golfo de Ancud. We took Canal Chiguao for a return to Puerto Quellón, a familiar safe anchorage amongst the fishing fleet. A friendly VHF conversation followed with the *Armada* station for our zarpe to Puerto Aguirre.

December 22nd, and time was as ever slipping away. We were far short of the Falklands for Christmas. Sailing is so much easier returning to familiar anchorages. One very special anchorage lay across the bay. We took a last walk about the town high street, smiles on our faces, frozen Turkey in our arms, and a carton of ice cream in the bag. We had as many fresh vegetables as we could carry. The dinghy was loaded down with Christmas tuck for our last run off the quayside.

Exiting Puerto Quellón all west, we were chased by Peale's dolphins on a short leg for the Bahia Anihue, as remote as we could imagine in the foothills of the Patagonian Andes. A family of friends waited for what would be a memorable Blighty Christmas turkey lunch in this far side of the world remoteness.

Chapter Eight
South for Fairway Lighthouse and a Right Turn for Magellan

Straights of Magellan, 53 and a half south.

"*The backdrop of the mountains keeps you focused on the mighty strength of nature.*"

"*We experienced a very special welcome of friendship in this very faraway place.*"

"*Glaciers of the ice field are rare moments of magic, this must be the most wild and beautiful place on earth.*"

The trip South from Puerto Montt had Christmas jumping out of the calendar. We could think of no more a spectacular adventure for such a special day than to visit the most remote chums we knew on Earth. We loaded up with the largest fresh turkey from Puerto Quellón.

December 24th and we were one day short of our destination. In a tiny remote anchorage, we conducted what has become a family tradition — "National Lampoons Christmas Vacation" movie for Christmas Eve.

Movie over and generator shut down, there was a stillness and silence in the darkness. The panorama was of volcanoes dressed with white snowy tops glistening in the darkness. Christmas magic above with a Milky Way sky of reindeer tracks.

Following the reindeer tracks for Christmas Day, we sailed into the beautiful Bahia Anihue at 43.52.30S 73.02.40W with dolphins listening to our Christmas carols. Sing it loud, "Thank you for the fish!"

There was one beautiful house in the bay. It was Pancho and his family's job to look after the house for its American owner. We joined forces for the special day. Our boat oven was too small, so Pancho and his family fired up the house oven to cook our prize turkey on land.

Christmas dinner began with marshmallows cooking ashore on the Patagonian campfire. The turkey was defrosting, but no Paxo. English Christmas traditions meet Patagonian remoteness. Big foot was absent, but our our Santa flag was flying on the mast. The GPS antenna was dressed with flashing reindeer and coloured lights. The church service was led by Morgause with a rousing rendition of *Hark the Herald Angels Sing*. The presents were placed under a native Chilean monkey tree. There will be none of those back in Blighty! Mr Potato Head gift for Caitland, and an electric skateboard for Morgause. There were no roads to use it on, but there would be tarmac in the Falklands.

Boxing Day brought a gusty 40-knot wind to blow away the Christmas wrapping paper. We liked Christmas in Anihue, but we had a mind to make a special destination for New Year. The target was Puerto Edén as the last Pacific outpost of Chile.

The tide and wind together beckoned us onward for a fast fjord run down the narrow waters of the mountains. We returned to a channel on the north of the uninhabited remote island of Filomena. These locations seemed like old friends.

There were seals on the rocks and shrieks of birds in the trees above a red rock shore fringed to the sky with a blanket of green fit for heaven. Was it really Boxing Day? With a two-point rope tie and the anchor towards the ocean, we sat, family together, on the bow awaiting a cloudy sunset on the far side of the world to say goodbye to Christmas.

Fresh eggs from the High Street of Aguirre

Awaking on the 27th with New Year in sight, there were beautiful islands all about us as we sailed from Filomena Island. We had grey showery days for southern hemisphere summer and a fabulous 180 degree all-south run. We sailed between a million small islands, each uninhabited and rich in green, untouched nature. After twisting turns into narrowing channels, we made Puerto Aguirre quayside wall for a tiny town of remote civilisation. Ropes were tight, and we were safe below the smart blue *Armada* office. It was a very sheltered bay twisting into the rolling hillsides.

There were coloured house roofs all about the place, water tumbling down the shell-covered pathways of the steep roads. No tarmac for the skateboard. The church and the school were closed for Christmas. We knocked on doors for fresh eggs from the store, bread from the bakery and even some cheese.

Each day onward with vigour, we had a mind to push hard for Puerto Edén and New Year's Eve.

From Aguirre, we had easy water on a long day hunting mooring buoys we couldn't find. Signposted to a new fish farm before Estero Balladares in Canal Pulluche, it appeared the buoys hadn't yet been placed. A tree-covered wooden plaque of boat names marked the old anchorage. It was shallow and rocky, but fresh river water flowed below Laurel Point, where ships of old anchored to take their water.

With our water tanks full, *Hollinsclough* moved up to anchor in more space at the head of the estuary. A tremendous view greeted us with fast tide runs for a perfectly timed arrival. The reverberation of echoes and the clatter of the anchor chain bounced all about the mountains, sounding our arrival. Hillsides of steamy trees are always a magical sight as they shed the evening damp. In the purple sunset, the trees turned brown into bracken at their tops. It was as calm as you like for easy sleepovers in the wilderness.

Beyond Laurel Point, we took a sneaky look at the open Pacific, the Gulf of Pain ahead. We dipped our toes in the ocean and checked the weather files. We sneaked back into a small estuary enclosed by a tall island in the Punta Refugio amphitheatre of monster mountains — a barren mountain-scape with bare rock, white limestone,

CHAPTER EIGHT: South For Fairway Lighthouse And A Right Turn For Magellan

steep and plain scary. This was our anchorage for the last evening of northern canals. Round the corner lay 200 miles of roaring forties ocean, a price to be paid before the southern canals of the Patagonian ice fields.

Tucked up at the very head of the Caleta Perrita with a four-point rope tie in a beautiful nook of green trees that turned the rock world back into a green paradise. We found a soft, gritty beach for a campfire to toast marshmallows. The backdrop of those mountains keeps you focused on the mighty strength of nature. It was an uneasy sleep before facing the Gulf of Pain ahead.

The strong winds were very favourable for our run south — 40-knot winds for the push to Rapier Lighthouse. Recording 47.6 on our wind dial, Windex was registering force 8 all along the north coast of Bay Anna Pink. We made a new record speed at 15.2 knots over ground. Wind with tide, we sped across the swell for a roller-coaster run. Our friend the *Evangelistas* heading north was being pasted. The *Evangelistas* ran at a much slower speed as she fought nature cutting into the waves. There were monsters at our back door, one pooping over the back to be sure the Christmas decorations were cleared from the cockpit.

The Gulf of Pain lived up to its reputation for a sleepless evening in the ocean as the tide turned against us. Nature was kind; she had delivered a late Christmas present to give us 190 miles of ground in 23 hours. Gulf of Pain behind us, New Year ahead and sheltered canals as far as the islands of Cape Horn.

Our trip up the southern canals had been a fight against wind and tide, but the trip down proposed to be a huge relief of favourable water in a good season of set weather. We had local knowledge and an understanding of the tide floods circling the Pacific-side islands.

Caleta Point Lay, the island of Little Wellington, and we were short of Puerto Edén by one day for New Year's Eve. It was still a celebration for our first anchorage in the southern canals of the Patagonian ice fields of the Andes Mountains.

New Year's Eve 2009

We found a magical anchorage below an enormous waterfall, green algae around the kelp with a mixture of ocean and fresh mountain water. Wild brown otters were in the bay, and a French yacht called *Savana*, which we tied alongside in a sort of six-point strap to shore. Team fry up for dinner, that's a sailing celebrations New Year's Eve with a glass of French red.

When you get proper international, one New Year's Eve is not enough!

We celebrated local time at eight p.m. to coordinate with France then nine p.m. for Blighty, and then midnight local for South America. We sounded hooters on the poop deck with new French chums Celine and Antoine — life on the ocean wave. You can't be waiting a whole year between celebrations — that would never do. Waiting for the midnight hour in the local location would never be the same again.

New Year's Day, *Savana* dug in waiting to push north. Her loss was our gain; the strong winds gave big speeds in the Angostura Island race track of tides. It was a Brands Hatch chicane run, sailing turns between poles, buoys and posts. The rock face smiled as local knowledge let us cruise closer than before across the undulating seabed of this fast track canal.

It was not too late for New Year celebrations as we arrived in Puerto Edén. Travelling south compared to north, we had covered ground at twice the speed. We had missed New Year's Eve in the tiny boardwalk town, but we were halfway in the leg from Puerto Montt to Puerto Williams and Cape Horn.

We experienced a very special welcome of friendship in this faraway town, but still no roads for the skateboard. New Year's Day dinner was with the *Armada* port captain, head of the *Carabineros* and their father and family. We presented them with a "Vive Chile" box loaded with McDonald's ketchup and mayonnaise. You don't get that down here!

We hand-pumped 400 litres of diesel from two old barrels.

Sat in Puerto Edén, we took breath for a few days.

Navimag ferry in the bay, Hollinsclough tied to the new queue at Puerto Edén.

CHAPTER EIGHT: South For Fairway Lighthouse And A Right Turn For Magellan

The majesty of the ice fields ahead. We would call at the Magellan Lighthouse, but there would be no mankind before the *Armada* base at Puerto Williams, Cape Horn.

Navimag ferry *Evangelistas* arrived back with some tourists on January 2nd, delighted to catch up with us after so many calls over the radio. We sang our best English Christmas carols for many smiles and a million waves. Crazy English abroad!

We took a few days' rest to explore the boardwalks and enjoy the wildlife in the southern hemisphere summer where colour is exploding from the green.

A new zarpe issued for a serious trip south to Puerto Williams, and an *Armada* team, happy to practise their English, brought the issued paperwork to the boat and stamped the all-important Teddy bear's passport.

Pushing south from Puerto Edén, we took a slight northern diversion. Dolphins danced in turquoise water before us as we made the run up Seno Eyre. We were hunting glaciers. Fifteen miles away and the magical white shadow of Pio XI Glacier came into view. It is the largest and most spectacular glacier of the Patagonian ice field. Our virginity of the vision of ice fields was long gone, but Pio XI still took our breath away, a magical field of time and elegance. Each field had its own character, tall, wide, aggressive and soft, sharp and dull, but all enchanted in vivid blue streaks of time measurement where rich colours marked the ice fronts.

We came about to port to anchor in Bahia Elizabeth for the evening. A two-point tie below the green trees of a beautifully sheltered nook anchorage protected us from all winds.

Pio XI Glacier

On Jan 5th, we took a day around the enormous Pio XI Glacier. It was true natural magic — blue ice, white water and mugs of steamy tea in hand. The Seno Eyre depths moved from 800m to 50m — must be as mountainous below as the view above. It was a snow-capped panorama, Mount Fitz Roy and Mount Piramide making near 10,000 feet straight out of the water before our eyes. Glaciers of the ice field are rare moments of magic; this must be the most wild and beautiful place on Earth.

Munch and crunch as slabs set sail from the mighty wall. There were crystal-clear cracks of sound before water swelled up from the explosions, the force rolling and tossing bergs uncomfortable to find the right way up. The small icebergs showing only their head and shoulders, hiding their under shapes for our imagination. It was time to take the dingy in close and how the enormity of the glacier grew before our eyes.

We returned to our easy anchorage, our evening onboard as flat as a floorboard hardly a few miles from the glacier in a nook of lush green trees. We took trips ashore to enjoy rare fungi plants and swamp nature, breathtaking views from a hillside moor that only the steps of ancient explorers enjoyed. We had a shoreline campfire to roast marshmallows and build a secret den from nature's ice debris around the corner.

Our dingy still sits well away from the face – it was enormous.

We rose early on January 6th for some extra mileage. An addiction to the ice was strong in our blood as we pushed south. Leaving the Pio XI canals, many dolphins were frolicking in our bow waves. The water was creamy, like milk, so there was very poor visibility for them and no salt — what on earth are they doing here?

With the wind favouring the week's run, our speeds were quick and consistently around eight knots of ground cover.

Lake District mountains were all about, Grasmere to Ullswater — how similar the variations were. We moved from grey slab 1,000-metre baron rock-scapes to thick green rolling forest hillsides. We left Canal Wide for Canal Concepcion.

CHAPTER EIGHT: South For Fairway Lighthouse And A Right Turn For Magellan

The glacier ice in buckets on the stern deck were our trophies, and dolphins approached for regular visits on the bow. Down below in even more contrast, it was simultaneous equations for the maths lesson. The English syllabus was interrupted by the arrival of a baby Peale's dolphin with its mum.

We left the aptly named Canal Wide for the narrow Schroeder's Passage. The epic views were restricted by the mist, but the depths were back to 2,000 feet and not quarter of a mile wide. We gasped at the depth sounding results. God's scissors must have cut a hole in the rock below.

Our anchorage for the evening was another return, Caleta Finte in Seno Fuentes at 50.21.80S 74.22.50W.

Last time it was the home of a huge jelly fish colony. Oddly, this time there was no sign of them. No rainbow of colours below, but on this visit, the mist above was gone and the view clear with breathtaking mountains, snow covered at the top. Like a dirty line on the bath, the tree line around us marked the step from below to above.

South, south and south some more.

Sixty miles for January 7th, we caught sight of a big cargo ship, the *Southern Ibis*, in Canal Sarmiento. The winds were with us again, and we had many good tides as well — the water was well in our favour. We were gaining more understanding of the tide flood. Water flow in the narrow canals was direction-focused by its pattern of entry between the outer islands. Watching the geography, judging the tide like the stream of a river, you could gain ground with a good choice of one side or another of the small islands. It was almost a two-way street on the tides.

We had a small view of the open Pacific as we tucked into Isla Evans.

The anchorage of Puerto Mayne was at 51.18.90S 74.05.00W.

The pilot book listed a small nook, but to our surprise, there was still more than enough room for the wind to turn us about a dozen times while we set a four-point rope tie. The girls dashed about in the motor dinghy. Morgause expert on the motor turns, Caitland would tie rope loops on the trees. When all were ready, we would commit the yacht in close and draw up the main ropes to the loops.

Safe as you like, the view in Puerto Mayne was a little baron. There was a lush green tree line, but it soon ended into mighty grey-white rounded boulders the size of Blighty mountains. Bedtimes when travelling are always good for celebrations, and 13,500 nautical miles on the log marked the evening's toast.

On January 8th, Faro Lighthouse was about 80 miles, so we chose to take a halfway run for a rest day. The weather was grey and wet, and after a quick team effort to leave our four-point tie in Puerto Mayne, we took breakfast on the water as a fast wind took us down Canal Sarmiento.

We sighted a tiny fishing boat hugging the shore of Carrington Island. There were many seals in the water, mini Loch Ness monsters jumping like giant black slugs from the waves. They say divers of old used seal skin before wet suits, so the Nessies must have been warm enough. Now in the land of the fifties latitudes, the temperatures were definitely on the way down.

We took a short cut between Taraba Island and Newton Island. It was a proper adventure run with five miles of nail-biting depth-sounder watching. We twisted down between a spider web of kelp and rocks. Isla Hunter reared up as we returned to the shipping lane of Canal Victoria. We made a port and starboard marker buoy and came about to the head of a waterfall tumbling from the mountain.

Our afternoon anchorage was lined with more kelp and rock islands for another breathtaking Patagonian wilderness. Anchor forward and two wide stern lines were backed up into the wind and tied beyond the shingle beach. We made steamy mugs of tea just in time for the evening rain.

Icy water danced above to play musical tunes on the sun canopy for a natural crescendo finale. Nature, no it's not over yet. The wind change, and the anchor slipped in the kelp. This meant an all-out team effort to re-anchor — who said it was an early evening? Dash, the tea had gone cold!

On Saturday January 9th, anchor and ropes were released, and we motored out of a Jurassic landscape — some of those views will haunt our memories forever. Otter Island by name and nature, these slimy, silky, slippery mammals were playing in the kelp shallows as we left.

At 52 south, the roaring forties were long gone as we sailed deeper into the screaming fifties. The land around us changing at a pace. Mountains lowering, white limestone-like rock and rich low green scrub. We passed an Oxford Circus of red and green buoys around Otter Island and then Piccadilly Circus for Shoal Passage and a view of the magnificent red rusting hulk of a giant cargo ship. There's a memento of mortality in these waters as the modern wreck signposted the route south into Canal Smyth. Afternoon sunshine in the summer grey rain was reflected on the Fairway Lighthouse.

The 50th parallel south, and a few days at the lighthouse.

Fairway Lighthouse made a most tremendous trophy for the most unusual anchorage. The lighthouse marks the eastern entry to the Straits of Magellan. It was only the local knowledge of a brief visit on the way up that made this anchorage possible. We knew we could actually moor to the landing rocks, and we knew it was well sheltered.

We ran ropes onto the rocks in the lee of the tiny island. Three giant dumper truck tyres fended us off. We scampered up the hill to meet the *Armada* keeper

CHAPTER EIGHT: South For Fairway Lighthouse And A Right Turn For Magellan

Pablo and his wife Cynthia. We were their first visitors ten months ago and likely their last as the year of duty was to end in ten days' time. What stories of Chile we had to tell.

Pablo set about a grand tour of the lighthouse, built 1920, it had two fine green Lister diesel generators, eight banks of six-volt Trojan batteries making twelve-volt circuits. Another set of two-volt 450-amp gel batteries lay in the foot of the small tower. Pablo gave us the gift of a lamp bulb — twelve-volt 100-watt halogen. What a prize! He showed us the spare emergency beacon and the systems balancing the charge for their home and the lamp. Life in the lighthouse was similar to life on the yacht.

What a view was to be had. Across the mighty Pacific, we could just about see Australia in the distance; it wasn't 10,000 miles away!

Cynthia baked fresh bread for our tea. There were scrambled eggs to make a real treat over soft butter, and even milk boiled coffee. Birds, mammals, and plant life were abundant and unaltered, undisturbed, unattended and almost unnoticed.

For January 10th, we were blessed with a soft wind change that gave us another day on the lighthouse rock. It was time for a rest and a grand walk about on the island of Fairway Lighthouse. Our *Armada* host Pablo called the weather forecast over the radio as we drank tea by his side.

Throw a dice and work out the probability of a tall-mast sailing ship going by while you stand on a remote rock.

We stood at Fairway Lighthouse in the evening sunshine to see Chilean *Armada* training ship the *Esmeralda* sail by. It was 300 feet long, had four masts with three sloops and more canvas than a world Scout Jamboree Camp.

We made a radio call to *Esmeralda* to wish her well. In reply, she informed us of her route. She was heading to Puerto Williams and Rio for the Bi-centenary Tall Ships Regatta. That was our route in reverse. What a small world in this faraway place.

The Monday morning rain of January 11th was fit for a car wash. Our spirits

Chilean Navy tall ships training vessel Esmeralda passing by Fairway Lighthouse.

were not dampened as we released the ropes and said goodbye to Fairway Lighthouse.

The sea was calm with no wind, and it gave us the opportunity to whizz around the entrance to Magellan on Polly Perkins. There was enough radio traffic to believe we were running about London.

The first radio call was to Fairway, reporting our departure and next ETA. After Oxford Circus, we turned down Haymarket to arrive at Trafalgar Square. The next radio call was to cargo ship *Jupiter* for a green to green pass. Then radio to *Sanpam*, a fuel cargo ship asking us for our intentions. Replying with great courtesy, he offered us a weather report on the Magellan Straits and took time to discuss the time of day. In very well spoken English, he wished us well on our travels.

Radio Felix, the lighthouse watch on the south side of the Magellan was next. He asked us to confirm our ship details, position, course and next port of call. After all that, he then made another radio call to practise his English.

For our return to the Straits of Magellan, the water was as flat as a floorboard — little wind and a neutral tide. We couldn't complain with eight knots on the dial.

Mountains of white rock lined the sides of our narrow world. The Peninsula of Cordova grew tall and then Playa Prada opened up to give us a way into the Patagonian door of rock. We had anchored here on the way up, so we chose to be more adventurous and twisted in to a tighter cut aptly called "Chica". It was like parking a car in the garage, only to find you can hardly open the doors. Four rope ties, snug as a bug, and we almost able to walk ashore. Scones were baked in the oven as kelp wrapped about our keel.

Snug in an anchorage marked as being sheltered.

CHAPTER EIGHT: South For Fairway Lighthouse And A Right Turn For Magellan

After an early morning in the dinghy to untie the four ropes on January 12th, we slipped out of the kelp and back into Magellan. Cargo ship *Parnassos*, 600 feet long, was our first radio call. Green to green, he gave us space on the sails. "Thank you very much indeed, sir." You don't get that in the English Channel. Tortuoso Passage for lunchtime, we came around Tilly Bay and 50 knots of wind came out of nowhere. We had a big push us for Ingles Passage. Is that a force nine?

We considered a return to Slocomb's Wigwam Island, but frighteningly accustomed to the 50-knot specials, we made starboard right turn — Deep South.

More south than we had ever been before.

Chapter Nine

Glacier Avenue, Puerto Williams & Cape Horn

55.5 South, Caleta Martial, we could smell Cape Horn!

"*Tierra del Fuego land of the Indian, Land of Big Foot...*"

"*A bucket of King crabs swopped for a bottle of red wine – that was dinner sorted!*"

"*Rounding the 'Orn...*"

With 180 on the dial, all south and more south than ever before, we could smell Cape Horn.

Out of Magellan into the narrows for Canal Barbara, 180 on the dial 53 south, it was time to get the thermals out. Water spray into the air… tornados? No, a pair of minke whales. They closed the bow by 20 metres for a fabulous display in the narrows. As we entered Shag Passage, there wasn't more than 200 metres of gap, and a late tide was pushing the might of the southern water into Magellan. We rounded Corner Point of Wet Island for a full-blown twisting whirlpool. A maelstrom of bath plug decent and southern side of the world, it was descending the wrong side of the twisting road.

Eight knots of tide against us, some sail and Polly Perkins the diesel motor full on. We were down to less than two knots ground speed. Large sea lions on the rocks watched all the fun. Cliff sides as steep as you like for floor-by-floor residency of shag cormorants. A million pairs delayed their fish dinner to see us safely through.

With the magnificence of a glacier ahead to end a long day, we turned right starboard into Nutland Bay — tree lined and very, very shallow. Kelp was strewn all about. A bow and stern two-point rope tie was listed in the pilot book. Keeping enough depth for the keel, it was very wide for a two-pint tie. Closer in, we would be aground. The girls were in the dinghy dashing about, rope laden with kelp for no easy knots in a fraught side wind. Morgause twisted and turned the dinghy motor to spit kelp from its propeller blade. We used the heavy sail winch motors to draw the rope, which was at one with kelp. We got it drum tight as we finally lined up side to the wind. Kelp lifted from the water was strewn down the rope like a shelf full of trophies. Morgause smiled. "Those rope loops are on strong trees tonight."

South of Magellan and farther south than we had ever sailed, it was a late bedtime, but finished tight as you like. Glaciers, gales, whirlpools and whales for a typical day out in Patagonia.

January 13th saw us back on the cruise-ship route.

Canal Barbara, deep south of Patagonian Chile, and we made an early start to bag a good morning flood tide and a long day on the water. Southern light comes early, helping those tides. It wasn't blistering hot, but we were getting more power into the solar panels than we had ever seen before. Gosh, the air must be clean down here. Despite the cold, we pasted up in sun cream to be sure.

Canal Barbara was a wide waterway of shoals, kelp and islands. The depth gauge was up and down like a roller coaster. The charts were out for island spotting, and they came along faster than train numbers. It was an astonishing variety of snow-capped monsters to soft, green, golf-course-like islands.

We left Canal Barbara to enter Canal Cockburn, and we were back on the cruise-ship route. Exposed more to the ocean swell, we took the first of the favourable tides for a land's end run around Isla Aguirre. This was a monster headland and gave us a door to Canal Brecknock, a tourist delight at sea level.

As we passed through the "creaking door", black dots shivered on the rock face. Sea lions abundant lay on slight ridges sheltered from the might of the ocean. Sleeping mountains, moulded by nature and time, gave way to a wild imagination. The extreme landscape was more scary than delightful. The mountains were weathered like rough sandpaper, granite contrasting white and grey. Each side of rock was bereft of all living things on the windward side. Rugged but curvy, deep crevices ran into lumpy mounds. Just plain rock? More like a lot of uneaten old buns scattered about a silver platter.

Canal Brecknock life was green to the south and grey to the north, nature's life is on a one-way street here. Trees like spider webs pinned back on the windward side and washed flat with the wind as they slither up the barren sides of the sheltered south.

We encountered the boggy seaweed at the shores of the sleeping mountains as we wound our way from Canal Cockburn to Canal Brecknock. It's like the scene from Harry Potter: "Relax and it will let you through!"

You feel the mountains closing in, peering into your journey, and with a flick of the wand, it lets you pass into another sleeping mountain chain bigger than the last. Standing behind like the eyes of the headmaster is the imposing presence of towering jagged white-capped mountains. Aloof and unimpressed by your presence, clouds of mist hover over their caps, maintaining a cloak of secrecy and reluctant to show the centuries of natural history.

Brecknock Island marked the end of the canal. Trees around its shore, green and red marker buoys define the bottom and a small anchorage to take our bow and stern ropes at 54.41.40S 71.32.70W.

It was a silent evening but for the steamer ducks in the stillness, but we then sighted another yacht — French sailboat *Orio* heading for the Isla Macias anchorage.

Hollinsclough was in the deep south of Patagonia, Tierra del Fuego — land of the Indian, land of Big Foot. What became apparent down here was how favourable the wind sits over the tides for our direction. Many yachts come the other way to experience the lower canals of South America, and those we passed were slogging into everything that was helping us. In an environment this extreme, help was at the top of the list; we could never imagine running these headlines against the prevailing winds.

CHAPTER NINE: Glacier Avenue, Puerto Williams & Cape Horn

January 14th started grey and damp with no sign of Paton Man, the Big Foot, but a large blue cruise liner passed our Brecknock Island anchorage, passing by too quickly for us to grab a wi-fi signal. Technology down here in 54 south country was unusual.

The old-fashioned system of Drake, Columbus and Darwin with the red magnetic pointer is hard to trust. The deviation from magnetic north scores around fifteen degrees of error. Can you imagine fifteen degrees out on the magic pointer back home? There was bottom of the world and North Pole bias like you cannot believe.

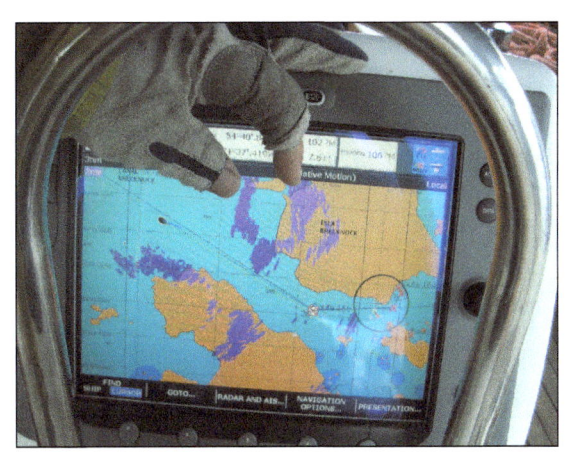

The purple radar should line up with the orange GPS land chart.

The GPS has been very accurate, but the chart datum is in error by between three quarters of a mile and two miles. NASA got men on the moon, but we have not accurately mapped southern Chile and Tierra del Fuego. What is going on in the world?

Meanwhile, the girls had just finished their KS3 algebra and started geometry — regular polygons, decagons and hexagons to keep us on our toes as we match the charts to the radar overlays and take a good look about to view the myriad of small islands.

Some Pacific swell was still pushing in, but all the tides favourable as we held almost dead east, minding that fifteen degree variation. We had rock banks and waterfalls on grand scales for a late lunch arrival at Canal O'Brien. We made an *Armada* VHF check-in with Timbales but got no answer in Blighty English. An *Armada* patrol vessel answered on their behalf. They advised us that their AIS was broken and the radio operator at Timbales was unable to speak English. What kind of an international VHF 16 watch is that?

The grey rain rose with a 40-knot push of wind. Chair Island roared up in the sunshine for a spectacular Jurassic Park look of an island anchorage bang in the middle of Canal O'Brien. We snuck about the island to come into the east bay out of that wind and up into a tiny nook called Caleta Cushion at 54.54.00S 70.00.00W.

There's a name for a soft landing Caleta Cushion — tiny beach but three drops of the anchor to get a grip through the kelp. With the anchor tight, we swung perilously close to the rock sides of the tiny bay, fending off to cushion the rocks. The girls were busy at all speed to connect their rope loops to the four ropes attached to the yacht cleats. Every corner of *Hollinsclough* was knotted in a giant spider web. Ropes tight, we had the safest spot in Patagonia as the winds outside tore down the twisted mountain passage. With so much wind so close above this tiny bay, we had full electricity with the wind generator blades spinning above.

As the morning sunlight shone on Friday January 15th, we released the spider web ropes and raised anchor from Chair Island to explore the Avenue of Glaciers — Brazo Noroeste, Canal Beagle. Snowy mountains pushing 10,000 feet and glaciers to the dozen gave us an ice day to remember at 54 degrees south. Our addiction to the beauty of ice was well fuelled.

The first glaciers came in a three like a run of buses. Romanche, Garibaldi and Pia. In the Avenue of the Glaciers, we headed for the most beautiful. Three glaciers merge into one at Seno Garibaldi. Mountains ceremonially stood to attention either side with tears streaming down their face. Cascading waterfalls of fresh melted snow covered the hillsides. Some waterfalls just trickled happily down to the sea, while others forced their energy into the ice flow. The stronger pushed through trees and over rocks to create a river plunging deep into the surface of the sea.

Large, healthy green trees grew in comfort hidden from the tortuous westerly winds as seals played amongst the ice flows. Small islands hugged close to the shorelines made up of rock falls from the mountain sides. A bay gave way to still water, and we anchored, gripping the soft mud floor. Here we proudly wrote our name on stone along with other intrepid explorers — *Allegro* 2002, *Vilja Mead* 1998, *Pasanoca* 1978 and now a decade into the second millennia, *Hollinsclough* GB 2010.

Pushing onwards toward the ice, we crushed into the ice flow. The water changed from turquoise green to crystal clear. All shapes and all sizes of ice came to nudge our hull. Hand-steering slowly, we glided through the small bergs and edged past the large ones with care, the waterfall crescendo singing to us as we went by. As each berg passed so another took its place until a barrier of surface ice said we could go no farther.

Hollinsclough sat still in the ice as we celebrated with lunch, at one with the glacier, still in a light reflecting rainbow of colours from ice as old as the dinosaurs.

Unhappy to leave this mountainous place, we accepted nature's rebuff as the surface ice pushed us away from the glacier front. Polly Perkins motor on, we turned south to leave, the ice flow swimming along with us now like children going out to play. Nature channelled us to the narrow exit over the sand bar, and we returned into the Beagle Canal bound east with enormous smiles on our chilled faces for the grandeur of nature at its finest.

One more mountain and we were at the opening for the glacier Ventisquero Romanche — tempting us, beckoning us, sun shining on the entrance. Couldn't resist. There was four metres of clearance on the moraine deposit bar. Clearing the narrow entrance, we found it flat calm. The sea lapped the shores of the giant's feet in a fight between saltwater and fresh melt ice.

We sat under the huge towering presence of the black pearl mountains. Icy rock faces glistering in the sunlight rays pointed the way along the sea to the head of the glacier. Inspiring heights mixed snow caps with clouds whilst dazzling you with their brilliance. These sights were so breathtaking that the steamy air from our lungs cleared the ice from our eyes.

Nuzzling slowly to enter the ice fields, the amphitheatre of three more glacier runs had showered the sea with bergs. These icebergs, trapped by the narrow entry, don't flow away; they are like children holding onto their mother's apron strings. Huddled together, protecting each other from the winds, they swirl and dance in a circle, happy to be near the glacier face and almost afraid to leave.

Hollinsclough coughed and choked as each berg thumped into the hull until the bergs accepted the presence of this foreigner. In the glacier front, huge rocks enclosed by ice peer at you as they wait for escape. Ancient soil deposits squirm and disperse little by little through the melting holes. Mountainous ice rocks too big to float at will have been pushed into the shores, creating a micro colony of icebergs. They swirl around pools of colour as the different shapes of ice float round and round.

Our minds twisted in astonishment as we left for the open canal.

Brazo Noroeste, claimed as Glacier Avenue, really is a must-visit location.

Twelve glaciers, sighted hanging from the Himalayan-like snow monster mountains, all floated by the windows of the yacht, ice holding secrets of the past for intrepid explorers of the future. Awesome!

Moving just a little east, the canal filled with turquoise melt water from the Hollandia and Italia Glaciers. Clouds from above touched the blue ice below. The blue ice broke to a giant grey hanging rock and down plunged a white-water rapid. It was like a cold tap full on in a steamy bathtub. The turmoil of heat and water rolled us about in ancient dinosaur swells. Family together, we looked at each other — how wickedly Jurassic park is that? We bagged a bucket full for the whisky.

Amongst all this ice magic, Gordon Island came starboard side — a mountaintop at 9,000 feet not metres away. It seemed like a lift from sea level to the top floor by nature, climbing the sky steeper than an express lift in Canary Wharf back in Blighty's London docklands.

Glacier Avenue, but the girls still had jobs on the rope work.

Glacier Avenue was a long way from Docklands and the River Thames. Two more hanging glaciers followed, almost less significant in the excitement.

We focused on the *Armada* station at Yamana. Caitland called in Spanish and they answered — magic! Don't be trying English on the VHF radio down here.

Bedtime dreams of dinosaur ice to make Caleta Olla — translation is cooking pot — perfect for a steamy spaghetti dinner with dreams of ice and a nightcap of glacier-ice chilled whisky in deep south Patagonia.

Saturday January 16th was a day with no rain in deep south Patagonia. Rising to the sun gleaming through the windows and not a whiff of breeze, we found neighbours for breakfast. A catamaran had joined us in the bay: *SV James McDust* heading north "the easy route" they said. Every tide and wind had driven us down — don't believe this is not a one-way street.

The sky matched the land. White patches of fluffy clouds reflected in the white patches of snow on the hilltops. Even the waterfalls make no noise as they silently followed their course. The contrast of Glacier Avenue to rolling hills was startling.

It was time to push ever east and try for Puerto Williams. If it were only a large naval base, it was still city lights, and it had a signpost for Cape Horn.

Beagle Channel gave way to the glaciers, but it was still lined

A summer's evening

CHAPTER NINE: *Glacier Avenue, Puerto Williams & Cape Horn*

by the white Himalayan-like mountains — Everest, K2 and Mont Blanc all around us. Caitland played it loud in Spanish. She practised the last of linguistics talking to the *Armada* station Yamana over the radio as the VHF range slipped away.

Fishing ship *Cisneverde* emerged around the first bend of the canal. AIS red to red, even the fishing boats were big down here.

We had a late lunch view of Ushuaia, a town of 50,000 people, 54-55 south. Ushuaia to the north is Argentinean, and our papers didn't permit us to enter Argentina without checking out of Chilean water on the south of the canal at Puerto Williams.

Ushuaia made for a magical view of city lights clinging to the edge of the water. The city sat sheltered under a crescent of enormous rocky snow tops. The place looked like Reykjavík, Iceland, and we chatted about happy memories of the northern folk.

It was time to change the VHF radio channel to 12 and English as best you could hear this far south. Argentinean Ushuaia port control: "Do you need anything?"

"Many thanks. We are all well and heading Puerto Williams."

Can't argue with that as we pick our way across the northern shallows of Canal Murray.

The Chilean *Armada*, not to be missed out, came alongside for a chat. In a small Rodman patrol boat, they were on a search and rescue for some local kayaks in the Beagle Canal. Kayaks? That's adventure for real!

Puerto Williams – 54.56.10S 67.37.10W

We arrived off Puerto Williams at five p.m. with 14,000 nautical miles on the log. It was basically a Chilean *Armada* base with aerials and watchtowers and a pair of PSG ships parked side by side on the pier.

The cruise ship anchorage buoys were both empty. *Hollinsclough* came to alongside the *Armada* patrol vessel PS16. We took a breath and Caitland called in Spanish for permission to enter the yacht club channel.

The yacht club was an old munitions ship that it lay in the mud, its name *Micalvi*. We tied alongside the rusting 1920 timber frame for a new home hosting a dozen yachts. At rest, we had happy memories of present-day *PSG Micalvi VI*. It had been our lunch stop for the sailing regatta in Puerto Montt, where Morgause had been the dinghy champion of the ocean. With so many glaciers, the warmth of mid-Pacific Chile seemed as far away from our new yacht club as dinosaur times.

The Micalvi Yacht Club bar opened at nine every day — 54 south for travellers on the far side of the world to meet and great. "Micalvi Yacht Club — the most southern yacht club in the world."

For our first visit, we had the company of Austrians, French, Dutch, German, Polish, Swiss, Chilean and three Falklands schoolteachers. Cathy, Jane and Pamela were on site to check the girls' homework. It was like Pete's Eats of Llanberis, Snowdonia, for climbers but with a sailing theme for travellers of the ocean. Pisco Sour, Chilean tequila juice, settling morning hangovers.

Micalvi Yacht Club, the most southern yacht club in the world.

Sunday service was held at the *Armada* church with officers in best royal blue. Getting more used to the Spanish, we kept up with the sermon, but better still were the chocolates given out as you said goodbye to the vicar.

Diesel would have to wait until Monday morning, but there was ice cream. The local *Armada* stores had fresh bread and pizza. The whole place was astonishingly warm. With grey, snow-capped mountains in the distance, you could imagine Heidi was nearby.

A Monday morning run to the *Armada* dock had our diesel full to bursting with the help of an *Armada* delivery truck. We were treated to a grand tour of *Armada* ship *PSG Isaza*. It had a pair of V12 Caterpillars the size of a house and a gen set bigger than a car.

Next quay lay a stealth ship set for evening patrols.

Armada tour over, we motored back to the yacht club — it was time to do the paperwork. Endless bother, as two of our passports were over their 90-day

Micalvi pontoon space was tight, and most yachts had a travelled look about them.

CHAPTER NINE: *Glacier Avenue, Puerto Williams & Cape Horn*

Chilean stealth boat

visas by twelve days. The PDI police threatened to arrest us, took our passports away and left.

Cavalry arrived in the form of English consulate John Kenyon — that's the Foreign Office service at its best. We had first met John in Puerto Montt, and what a relief for him to be sailing the canals and available for our rescue. Still, we had to wait two days for a visa clearance — how crazy is that?

The wind down here is very changeable, and there was nothing to let us out for a few more days, so we began some tourism.

On walks into the mountains, rivers charged down and beaver dams defended the rush for miles. We made it to the top of flag hill, the pole short and tethered four ways against the wind.

Flag Hill on a calm day

We visited a small but special museum of ancient Tierra del Fuego Indian life with portraits of Fitzroy, Darwin and Columbus, who had all been here before us.

Still without passports, Friday arrived and then everything closed for the weekend. We made daily visits to Yvonne at the government office, and each time we arrived, she said maybe later! Mañana, Mañana!

We had a daily change of neighbours as charter yachts working their trade delivered ever more explorers to the yacht club. You could fly into Ushuaia, meet a charter yacht, and inside a week — weather permitting — sail the loop of islands to round Cape Horn.

Peter and Rob arrived from Liverpool bound for Cape Horn, winds gusting at 80 for tales of true sailing. *Star Princess* lay in the outer bay for an hour — 1,000 feet of ten-deck cruise liner in this far faraway place. It had a TV screen so big that we watched the outdoor screening of the BFG (Big Friendly Giant) from our yacht.

Still waiting for paperwork in the most southern town on Earth, it was time for a full day's trek. Family together, we got a lift in the local builder's van to the end of the road, where we found dense woodland and a mud path for an 1,800-foot climb. The tree line came to an end, and it was open gravel as far as you could see. In a windswept world, it was upwards and onwards to the snow line and a fantastic view.

Every evening we enjoyed drinks at the southern-most yacht club in the world.

Cool but not Chile, crazy stories, adventures of heroes and glorious boats, there was not a normal person to be found. Anotonia, the Polish tri-athlete who cycled, skied and kayaked from north to south Chile without a zarpe! The Whitbread 1988 winning 60-foot race boat now used for French Antarctica explorers, the race boat had been converted and it had a large kitchen! The 40-foot *Ovni*, which takes ten passengers to Cape Horn but only has six beds! The ferry to Argentina that is just a dinghy with a roof!

This far-side-of-the-world yacht club was no normal place. We left a Royal Channel Island Yacht Club pennant on the board. Like the charter yachts, it was time to take a stab at Cape Horn.

It had taken ten days of paperwork to leave Chile. Stamped and done, it was west. A short 30-mile day down Mackinlay Passage. We called Snips station on the VHF and made a right turn to pass 55 south at 14.30 on the 28th January. All south for Puerto Toro in a soft wind run.

A very sheltered bay, big new pier to tie alongside, water on tap and 16A electricity, all free of charge and not 50 miles north of Cape Horn at 55.04.90S 67.04.50W.

CHAPTER NINE: Glacier Avenue, Puerto Williams & Cape Horn

Puerto Toro is one of the earliest settlements of Tierra del Fuego, but gold prospectors have been replaced by fields of buttercups. Today, there is still no tarmac, maybe 50 tin houses, but a good number of satellite dishes. There was an *Armada* office at the top of the hill to get our zarpe diary stamps. The weather forecasts were printed on paper in best Blighty English. A bucket of king crabs was swapped for a bottle of red wine, and that was dinner sorted.

Puerto Toro and a hook up for the electricity

Electricity ended at nine when the diesel town generator shut down for the evening.

Ten miles north of the Horn, this was life in the islands at the end of the Earth.

Unplugging the electricity line, we left Puerto Toro quayside by Gore Passage. Forty miles more south across the Bahia Nassau, which was very open 55 south water. We had a force six beam gusting forties for a flyer on the sails.

The water was shallow by local standards at a few hundred feet. Swell, wind and chop was all confused. The run was bumpier in the mind for the reality of Cape Horn on the horizon. *Armada* patrol launch P1603 did a fly-by while heading back to Puerto Williams to check all was well. Albatrosses down here were a size we had never seen before.

More chums in the air, on approach to Wollaston Islands, tourists took our photos. There were eight of them all tucked up in their big red Bell helicopter CCCGM — the places you have to be sure your hair is tidy!

The wind fell away between the islands as we entered Canal Bravo with a dozen dolphins in escort. Small

King crab for dinner

147

derelict houses sat in the coves as we swung around a few miles of the channels to Herschel Island and anchored in Caleta Martial at 55.49.30S 67.17.90W.

We anchored on 100 metres of chain in six metres of water, long rope on the stern to the wide soft yellow sand beach — just can't be too sure down here. A shallow river streamed into the sand and a giant king penguin stood tall to great us. Bless, he was all alone with low-lying land all about.

There was a shallow tree line where just a little wood defended against the wind. Beyond lay open windswept plains of lichen and bracken, which was in full combat with the elements of the screaming 50 winds.

We settled to bedtime not twelve miles north of Cape Horn.

A sunshine breakfast followed with more guests in the form of leopard seals on the beach. Back on land for photos of the king penguin.

Paparazzi penguin shoot done, we took luck in the sunshine and edged out north of Herschel Island for six soft miles to Oriental Passage, three knots of tide against us on water that was mirror flat. We had our first view of Cabo de Hornos Island, two sharp conical horn rock hills to its west. We checked in on the VHF with *Armada* radio, frightened to commit the run for Cape Horn, like we hadn't paid in gale force penalty, we first chose Isla Hermite and the tiny anchorage of Puerto Maxwell.

Maxwell is a small cove with a crescent of rocks and kelp to enclose the ocean. Inside it was as flat as a floorboard. Mirror water, we could wave at ourselves in the stillness of the reflection.

Are we really 55 south? We dug the anchor in amongst rock and reversed up, the girls had rope loops and two shore lines ready for the winch. We were fit to survive a tornado.

The afternoon sunshine baked down from a wild turquoise blue sky. The solar panels were charging the dials into the red. It was a sky the like we had never seen. It was so clear, not an aeroplane trail to be found, but a wild clarity in the colour of the air that assured you this location is the end of the world.

With the kindness of the weather, we were privileged with the opportunity to walk ashore and take the elevator 300 metres upwards. Mulch and squelch on a ground of moss, lichen and bracken, soft on the feet, there was half a path marked by bleached white rock. The view from above was breathtaking — volcano-like granny teeth from ocean to sky in every direction. It was soul touching remoteness and some wicked photos.

A full moon in a cloudy sky, so no galaxy vistas, and the Aurora Australis still evaded us.

CHAPTER NINE: Glacier Avenue, Puerto Williams & Cape Horn

The 55th parallel south and flat calm

February 1st 2010

It was time to become *"Orners with the bottom button undone."*

Ropes away from the fantastic anchorage of Caleta Maxwell, we made a starboard turn all south for a tiny narrow passage between the rocks. It was the scare of our lives when we caught a rock in 20 metres of water. There was no a sign of it on the charts. Maybe it was the mast head of a sunken yacht? We checked the bilges and all was well.

We set the sails for a soft short run and recovered with a stiff cup of tea, as Blighty folk do. We turned off the autopilot and took turns hand-steering for the Horn.

We sighted our first cruise ship of the day, the 500-foot *Le Diamant*, at about two p.m. It was resplendent in the sunshine between us and the jagged rock Island of Hall. We bagged 56 south on the dial by two-thirty — that's Antarctic territory to you land folk!

We passed Cabo Hornos light on a transit from west to east at 13.30 Febuary 1st — "Rounding the "Orn!"

Caitland and Morgause had the yacht and took turns to hand-steer Cape Horn.

Hand steerage for Cape Horn

Tracey and I just took in the view of wild rolling flat green mountains with battle-scarred rock faces — not a tree in sight.

There were many holes, like a lunar landscape, from impacts of the weather, but we had scary-clear turquoise blue skies. Streaks of whitewash-painted cloud splashed over the sky to highlight a distance farther than sight allowed.

Never mind the view. Hooters, cheers and thanks to God in a resplendent action of all things sailing. It was a jubilant moment to round the Horn hand-steering on the sail. We had earned the right to wear a navy blazer with the bottom button undone.

The imagination that we were all alone was soon broken.

Three other sail yachts were sighted as we cleared Espolon Point. Then, at the southern tip of Horn Island, we found our second cruise liner of the day. Six hundred feet of the *Delphin* lay at anchor below the rocky landing beach of the Horn light with a tennis court on the top deck! Cape Horn, what a racket was going on.

We think the tourists took more photos of us launching our dinghy in the mighty swell than they did of Cape Horn. After the surf beaches we have fought, it seemed a doddle, but it no doubt looked perilous to the bystanders.

Family Moto: "Every day is an adventure." Swell, surf and stones to make the shore and climb the wooden steps. At the top of the hill sits a big red Bell Huey chopper ready for rescue duties. A team of Chilean marines were on bomb disposal and mine clearance.

From here, it would be bottom blazer buttons undone.

CHAPTER NINE: Glacier Avenue, Puerto Williams & Cape Horn

Statue of the albatross at Cape Horn

"Stay on the path!" cried Morgause. What no Argentineans!

The famous metal statue of the Wandering Albatross was epic. What a setting, erected for the memory of sailors who have lost their lives in this water. Photos, photos and more photos. Morgause donned her Chilean Girl Guide knecker in front of the signboards with the iconic albatross sculpture behind her.

Then came a visit to the lighthouse — 1902 and all.

We climbed the stairs, covered in sailing burgee flags, and saluted all those who have helped us arrive here. The *Armada* navy station stamped our passports — takes a whole page, and it was the coolest print in book.

They stamped the yacht's log, the girls stamped their Girl Guide records, and then we all stamped ourselves — a tattooed family for celebration at the Horn.

In many touching moments of travel, some places are special, but this one down here seemed a big one and definitely touched our hearts. Morgause, an eleven-year-old, was the youngest 'Orner in the family.

The *Armada* sergeant on his year-long tour of duty said westerly winds often made 90 knots but blows from the north rarely go over 25. Time to run.

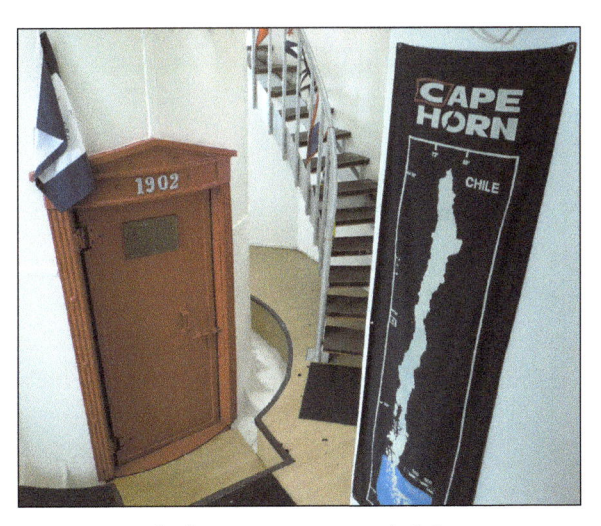

Inside the 1902 Cape Horn lighthouse

Motor on, we fled headlong into fifteen knots of northerly for a ten-mile dash back into the protection of the small islands.

Our evening anchorage back at Caleta Martial, Herschel Island was still 55 south. The swell on the rocks were still like thunder when the third set lands. Some sort of micro climate kept this whole place surprisingly warm. When the sun shines, it scorches through the lack of ozone and bakes you in the southern summer of Antarctic wind. Keep blasting your ozone spray cans, as it's saving our heating fuel on the boat!

Islands at the End of the World

At 55 south, recovering from our Horn trip, the king penguin we had named Herbert sat motionless on the beach. We lay soft on the ship mooring buoy of Caleta Martial. Dong, Dong, Doonggg! We banged on the buoy as we lay due to the lack of wind on the tide. Who said the Horn was a rough and wild place?

Tracey cleared the sink with the bilge bucket and threw the dirty water overboard. With the dirty water went two sets of crockery. Whoops! That made the boat a little lighter.

The sea and the land blended together harmoniously, and the light of a full moon showed the wilderness. Still no Aurora Australis in the short summer darkness of the evening.

There may have been no coloured light show in the dark night sky, but now we could smell the Falklands.

From Cape Horn, a three-day run avoiding the big Pacific flood tides exists in the open ocean.

We needed three or four days of weather window. Bizarre as you like, only two-day gaps were coming up. Mighty westerlies were all turned by unusual depressions. They held for a few days, then broke and turned.

We took a day out to put the motor on for Canal Bravo and took a last view of the southern islands of Cape Horn. Maxwell and Martial anchorages were resplendent in sunshine and soft wind. It was hard to believe this was the end of the world. Fantasy or reality, the vision before our eyes merged between the two.

We were all touched by the geography of the place. A barren wilderness of low moss-covered rock that protrudes like a whale breaching. There was an eerie feeling of loneliness, like fire embers burning out. The feeling that at any moment creatures

CHAPTER NINE: *Glacier Avenue, Puerto Williams & Cape Horn*

unknown would fly past to bid you a fond farewell. We may have the right of passage in joining those who have "rounded the Horn", but the excitement dulled in recognition of just how far away we were.

We called Wollaston radio to check in our position with the *Armada*.

Windows of weather not there for a direct run to the Falklands, we hatched a new plan. We would knock a day out of the run by heading 40 miles north to the *Armada* Island station at Lennox.

That would give us the strong tides of the Le Maire Strait but bring the Falkland Islands more west.

The weather could change and pin us in? Surely not!

Chapter Ten
Cape Horn bound for the Falklands

Albatros statue Cape Horn.

"*Cape Horn had taught us caution.*"

"*Nature had delivered us a calmness in the centre of a mighty gale.*"

"*The southern hemisphere sunrise was a soft grey affair…*"

Sitting at anchorage a dozen miles north of Cape Horn, our adrenalin was running high. In contrast to the geography, we were bathed in February sunshine.

Ahead lay the Falklands to restock Blighty rice pudding and tins of baked beans. Beyond that lay South Africa. With Cape Town in sight, there was much stability of the master plan in our minds. The girls would fly home to school for September. We had missed Christmas in the Falklands, but time and seasonal wind patterns were on our side with some margin to spare. From Cape Town, we would sail north on the trades for Ascension Island and take a look at the Panama Canal for a run back into the Pacific. If Somali politics were settled, that would clear a circumnavigation route, and we could have another look to see if the world really is round.

Caitland used her best Spanish to call Wollaston radio and clear our position with the *Armada*. She explained our zarpe papers and outlined our Chilean waters sailing plan to the Falklands.

The Falklands should have been an easy wind run. The winds generally howl across in very set patterns, but the season had brought a great deal of unstable wind turns as low pressures twisted the air passing over our heads.

Struggling to get a set pattern in the wind, we chose to make a day to the north and dog-leg the route to shorten the open sea run to the Falklands. We would run 40 miles to the *Armada* island station at Lennox across Bahia Nassau, then change course to fight the tides in the Le Maire Strait and come to the Falkland Islands more westerly.

Bahia Nassau had been a roller-coaster run down, but the weather gods liked the new plan and let us slip smoothly across. So smooth that it seemed nature agreed with the decision and English school lessons took place aboard. Smooth waves, cups of steamy tea, Pot Noodle lunch and a splendid late afternoon arrival at Isla Lennox — kelp strands all about. We took a cautious run north of the rock island Isla Ormeno to be sure.

Even the anchorage was easy; we lay 50 metres of chain and drew back tight, securing the anchor solid in a sandy bottom.

We spoke on the VHF to the *Armada* in their small blue hut. The quayside under repair, and some new houses were being built of smart clean timber. Anchor down in a long shallow sandy bay and we were ready for an early dinner at 55.17.70S 66.50.40W on February 4th.

As we awoke the following morning, the large blue German Ketch *Vanhee* was anchored to our port side. We left first for an early start bound for the Isla de los Estados on a 90-mile run.

The morning water was fairly neutral on the tide and there was a good wind behind us. As the tide turned, it became tide against wind and chopped up into a mini monster. The sea was white and frothy as we cleared the north Isla Nuevo. The mini monster grew and the swell rose to five metres high.

A large red fishing ship slowly pushed back the water trawling ahead into the building wind. The wind gauge topped out at 45.2 knots for a forecast that had been fifteen to twenties.

There was a full gale on the stern by lunchtime. As the tide returned to neutral, we achieved a best ground speed at 11.8 with most of the time over ten knots. Motorboat speeds, we were hammering along, but the eggs were scrambled.

Trained in the Gulf of Pain, life was OK, but it was no way to start an open sea run with so many small anchorages around us. Cape Horn had taught us caution, and let's not forget, this is 55 south country. The weather forecast windows of set wind remained short, and this change stamped approval on waiting another day.

Appropriately placed south of Bahia Sloggett with 40 miles on the log, we pulled in.

An easier option was the remote Argentinean anchorage of Bahia Aguirre on the north side of the channel, but we didn't have a permit for the Argentinean water. We had cleared Chilean paperwork and attained exit stamps from Puerto Williams, but this was the first time we had left the Chilean side of the water.

It was all stations to drop the Chilean courtesy pennant from the starboard mast stay. Say goodbye to Chile and enter Argentina for an evening of tango and Buenos Aires football memories.

Argentinean colours on the mast, it was all about north up and into the bay. We gave a wide berth to Kinnaird Point, rocks and kelp everywhere. Then we made a little push back into Punta Pique to pass a large ship buoy bouncing about in the outer swell. The big grey beast was a little far out for protection, so we made for the grey shingle beach and dropped the anchor chain in about five metres of water at half the tide. It dug in well, so we locked the snatch rope onto the chain and we took in the view at 54.55.70S 65.58.30W.

There were trees all about on big green hillsides. For so few miles, this was so much more homely than the barren lands of Cape Horn. Sunshine and sea spray provided some window cleaning duties. Hail and snowstorms followed, and that agreed with the decision to turn for the shore.

A variety of old derelict houses and shacks stood along the beach, but otherwise the place was devoid of all but us. We changed the VHF from Chilean 14 to

CHAPTER TEN: Cape Horn Bound For The Falklands

Argentinean 12! We were watching radio channel sixteen, and fingers crossed, the Argentineans had no patrol boat about to kick us out on our Chilean paperwork.

Cape Horn adventures filled our evening dreams. The window cleaning had been worthwhile, as we awoke to sunshine pouring through the glass for a beautiful morning.

Wind set back to pattern, so we left the anchorage in Bahia Aguirre on the Argentinean side of the Beagle Canal for another stab at the Falklands.

Upside down tides to work out, the falling Pacific rushed out, and on the other side of the straits, the Atlantic push was a half tide forward.

The best dash calculations said to leave at nine-thirty, and a stern wind agreed. The water gave us a high of 11.4 ground speed — this was rocket motor speed. We cleared into the Le Maire Strait proper for a very good run.

The wind was softening after a top gust of 38 and a swell that rolled out like the waves of the *Titanic*. Slowing down by afternoon on approach to Staten Island — Isla de los Estados — we sighted a sloop crashing and bashing back in. We got on the VHF to say hello. It was single-hander Bob on *Sylph VI*. He had been travelling down the Argentinean coast and was heading for the very fjords and adventures we had left behind.

"Give regards to my chums on *Flying Penguin* in the Falklands," said Bob.

Staten Island stood before us, four mighty bare white rock mountains sat in the sky like the humps of two camels. Memories of Fernando de Noronah and Robinson Crusoe Island were first to mind. The barren emptiness of this place brought a scary side to the view. As the panorama grew close, we had taken our eyes off the speed dial.

The tides weren't going to let us pass. Nature's rocket motor of Pacific flood was out of fuel. We were staggered to watch the ground speed gauge fall below two knots. Crawling like a baby, we continued to be two hours away from our San Antonio GPS target marker for another four hours. Sailing at a pace on a good wind, we were near stationary. Close in, we found a little lee and got ourselves tight to the headland. The tide was on the turn, and then we were back to eight knots for the same sail.

The tide was in a local cycle and clearly turning more than nature intended. The majesty of those mountains outshone the might of the next tide, and we sneaked into the Puerto Hoppner ship anchorage to await another cycle.

Wide and well sheltered, it was splendid relief after a long day of three half tides. It was a natural amphitheatre of plain scary proportions even by Patagonian standards. It was the apocalyptical bareness of the rock. Never inhabited and no green foliage, it was just a rocky summit string-like island of biblical heights jaggedly towering into the sky.

There was a building evening wind outside the ship anchorage. We enjoyed the anchored calm to tuck into dinner and added the luxury of a desert. The pilot book logged an inner yacht anchorage. That was good enough to signpost a splendid evening's sleep.

Without the Patagonian training, it's unlikely we would have attempted this entry. We traversed the tiny gorge entrance to the inner lagoon. The entry was hidden in its own shadows at the back of the outer anchorage.

A metre spare on the side, it was so tight we put fenders out for the rocks. We breathed in to slip between walls so close we couldn't have pumped the toilets. Morgause was in the dinghy out ahead on kelp duty to guide us across the shallows.

The smooth waters went mirror-like inside, and our eyes were met by an unbelievable contrast. The wild mountains remained touching the sky, but at ground level, there was a waterfall-edged green lagoon. Outside desolation turned to nature in full bloom like we had walked into Shangri-La.

Inside Staten Island

It was a team effort for a multi-point spider's web rope tie before bedtime. Having seen the devastation of bareness outside, we were taking no chances. We slept silent and safe tucked up in Jurassic Park bliss.

Morning song: "All things Bright and Beautiful".

As ever, the unpredictability of the weather had changed once more, and a big low turned over the roof to whip up some huge winds in the wrong direction. The step into Staten Island had been great fortune; all we had to do was work off the calories of that extra pudding desert.

The wow-factor of the view was up there with the wind speed. We watched in awe as the upper waterfalls crashing down from the mountaintops and turned into an upward shower spray. The velocity of the wind was greater than gravity; it was a fire hose like spectacle in the sky. It was raining waterfall, thrown from one mountain ridge to the next.

Time out then, we would sit here for a day or two and do a few jobs — clean the fuel filters, balance the fuel tanks, burn our rubbish on a shore campfire and roast the marshmallows — and celebrated the majesty of the place.

What had been an easy entry gave no easy escape. The Velcro was strongly enforced, not by city lights this time but by the wind turns closing the Falklands run. As a few days went by, the wind outside built from 30 to 40 and then 50 knots. There was no regularity in the direction as two lows moved over our heads.

Even the tranquillity of our inner anchorage home was torn up by the growing wind. On the third evening in this little haven, a stern rope broke. We awoke safe but pointing in the wrong direction. The girls put a second stern line out in the waterfall rain — well done the youngsters.

Nature had delivered us a calmness in the centre of a mighty gale sweeping across of Cape Horn. We were truly privileged to be locked up inside Staten Island as carnage was ripping up the ocean outside.

By the fourth day, the wind finally settled westerly with a clear pattern. We sneaked out to the ship anchorage in the dinghy, but the ocean was still ripped up. The Le Maire Strait, the junction of Pacific and Atlantic, had the leftovers of a battle against the low, and a single lost day would do no harm.

While the seas outside settled, we rigged the spinnaker pole for a wide jib. Two days would see us on a downwind run to Stanley in the Falklands.

After a few hectic hours to clear our four-way rope tie, we were ready to leave the Puerto Hoppner inner anchorage.

The tiny inner lagoon had become a friendly home, but it said goodbye with a 30-knot williwaw gust from the mountains just as we cleared the rock gate exit. Five metres wide for four metres of boat, just a metre below, and wallop, we were in the outer anchorage. Sails set, swell building, and we were away.

It was scones and flapjack for a sensible sea-bound lunch back in the Atlantic swell. It was a good day on the sails, but the winds by sunset were too soft. Polly Perkins was fired up by six in the evening to fight the tide, which was still strong.

The wild sharp pointed peaks of Staten Island slipped into the evening sunshine. Seals skipped in the waves and albatross sailed the sky. What a joy to find that the land-effect rain was gone — no more waterfalls raining down on us.

In the deep South Atlantic, we sat around the dinner table, family together and dreaming of English excitement — Christmas post, fish and chips and rice pudding. We toasted Great Britain with our dishes of steamy tinned stew.

Seven sharp, dinner devoured, we saluted Margaret Thatcher as we entered the 200-mile fishing exclusion zone — it certainly felt like England. Clouds stole the southern hemisphere stars as they covered the Milky Way sky. The winds built, and we settled in for a fast run in the darkness. The AIS computer sighted a 350-foot fishing ship named the *Tai An* as she circled us, trawling the rich Antarctic waters of the exclusion zone.

The southern hemisphere sunrise was a soft grey affair, the deep orbit position of the Earth giving a long twilight wake-up call.

Our morning coffee at ten greeted three large grey Antarctic minke whales, a curious family who came up close for a look at their fellow travellers.

A soft-wind day, and 200 miles of the leg were down. It was a sunset watch for the first sight of land as we headed toward Bull Point East Falkland. Maybe it was shadows in the cloud, but the sight evaded us. The Jimmy Cornell guide said run north of Sea Lion Island, so we did. It may have been better offshore, but the prize of that first land sighting is a mighty goal not to be spared. Depths were down to around 70 metres, and we followed the depth line in the dark with the eyes of the radar to match chart to datum. We heard the sounds of the sea lions yelping, barking and splashing in the darkness of their nocturnal dinnertime. Thick kelp danced in the shadows of our swell. A little vibration told us we could have caught a rope or hooked an extra tough strand of the slippery kelp.

Stuck steadfastly on the 70-metre line, we eased up East Falkland, outside of Lively Island beyond Choiseul Sound, and made Bluff Cove for a five-thirty sunrise. Warmth in the rays, we were at 51 south, which was positively north by our standards. There was still no visible sign of land ten miles out and approaching Cape Pembroke. The morning had brought soft wind but thick cloud with rain holding fast to the land — welcome to England. We sat about the breakfast table talking of

Margaret Thatcher and days of war — what was the name of that reporter chap who went with the Royal Navy? No Harrier Jump Jets in this cloud surely.

We had the AIS VHF locked in and perfect English on the radio for port control — channel ten please. Wolf rock, Cape Pembroke, the narrows and into Stanley sound.

We came port left and tied up to a sort of ferry jetty called the Flypast berth. Within minutes, we were met by Fisheries officer Roy Summers. "Just a moment, we shall call the FIC jetty and arrange for a proper home in town. After all, you are British."

Port Stanley – 51.41.50S 57.51.30W

Alan from Customs stamped us in, and Blighty flag flying, we were ashore. No foreign customs papers, we stepped onto British soil to make a dash for fish and chips at the Woodbine Café. The owner was a Leeds United supporter go the Owls! Time for bed, we would explore this Blighty world the following day.

Chapter Eleven
Falklands Fun

Blue sky moorings. FIC Quay with the ship Pharos in Stanley, Falklands.

"*15,000 miles of a world tour and we were in England?*"

"*Warship HMS Clyde came into the Sound, anchored and saluted lost comrades…*"

"*Shackleton himself looked us in the eyes and beckoned us for adventure in the ice.*"

A fter nearly 15,000 miles of world tour on *Hollinsclough*, it was hard to believe we were in the Falklands.

It was February, and the plan remained clear: restock with English supplies and sail for Cape Town. From Cape Town, we could access the British Airways flights to Heathrow and easily see the girls fly back for the September school term in England.

For a change, time was very much on our side. There was a hint of South Georgia for some adventure en route and a fly by Tristan de Cunha. The route was all set in good seasonal wind patterns. There would even be a way home option. After Cape Town, the trade winds favoured St Helena and Ascension for Blighty. If Somali politics changed, we could take stock of our world route for a left turn into the Pacific via the Panama Canal.

With time on our side, we engulfed ourselves in everything Falklands. We came for a week, but stayed a month.

That first Monday set the timescales. There were four weeks of the school term left, and the girls adored the art, sport and formality of lessons, so the Velcro was laid and the dates set. Headmaster Mr Baldwin invited the girls to four weeks of school.

The girls, togged in their Repton school uniforms, climbed over the yacht rail, stood on the FIC jetty for muster and walked to school.

Days home for lunch, we visited the quayside chip van and but for the other side of the world, we could have been at home in England.

Blue mafia, it was the Girl Guide hut next to the sports complex. We were not the only visitors. It was a big hello to sixteen Adventure 100 Guides from England on their week visit. What kind of timing is that? Guides the world over, teatime and tales of travel.

There was a torch for Tracey to carry with the Girl Guides. The Commonwealth Games had come to the island, and there she was in the middle of a Rainbow pack.

We went to the store for a view of Waitrose, England, had fun with fresh bread and cereals for breakfast. There are no cows on the Falklands, so the milk is still UHT. We even had an Iceland pizza fresh out of its frozen

School uniforms for the southernmost classrooms

The Commonwealth torch with the Girl Guides

cardboard box. Schoolteacher Pam arrived, who we had met at Micalvi Yacht Club. She booked us in for a romantic Valentine's Day dinner at the Brassier.

Landrovers were everywhere, plus a red 38 bus from Islington and red telephone boxes. We said hello to the visitor centre lady, Joan, and she replied, *"Ahh my niece at the post office has a stack of mail waiting for you."*

"Camp"? what's that when it's at home? Everywhere outside Stanley is called "Camp". That's a town and country divide to you and I, like a London and Leeds north-south divide.

We soon found the DHL office and some engine spares sent out by Roy: two new high-pressure diesel pipe replacements and a spare lift pump for Polly Perkins the metal sail. There was a new burner for the cooking hob; it would be paradise in the kitchen.

We made a family visit to the hospital and got some GP appointments for a Blighty health check-up. It was a sort of town cottage hospital all the way 52 south on the National Health Service.

Shrove Tuesday and Ash Wednesday saw us join in with the congregation of Christ Church Cathedral, Stanley. There were many a newcomer to church, as many islanders are visitors on two-year work postings. Teachers, policemen, scientists or nurses amongst them, we weren't out of place. There was no regular organists for the big tubes, so a modern electronic keyboard with a silicon-chip memory played the hymns.

We met Sea Cadets from *T.S. Endurance SCC*, and we made a presentation to the cadets of our trip so far with naval curtsey flags and a quiz for how many nautical miles travelled. Morgause proudly wore Mummy's Boatswain Call (ships pipe). She was taught how to play it from Leading Cadet Murial.

The average wind speed in Stanley was recorded as a force six. Local knowledge of wind directions is to be heeded. Ulrich from the FIC quay told us to turn *Hollinsclough* about. When he tightened the rope straps on the jetty building roof, we knew he meant it. We moved our bow north in flat calm afternoon sunshine to face the coming winds.

CHAPTER ELEVEN: Falklands Fun

Piccalilli on our lunchtime sandwiches, ginger nuts and shortbread with the tea. Ready to re-stock the boat, we tested tinned meat, corned beef and jam brands to be sure we choose our favourites before the Argentinean blockade kicked in.

Weekends out of school focused on traditional Falklands' tourism — war, war and more war.

Our history of the Margaret Thatcher days began with a battlefield tour with Frank Leyland. He fitted us in between cruise-ship trips. The family was squashed together in a Japanese four-by-four — no Landrover space today. We headed all west from Stanley into the shadow of Mount Osborne, which meant two hours across open terrain with not a tree to be seen.

The low rolling soft clay ground was strewn with volcanic rock strata from the time of the dinosaurs. There were minefield signs at the gravel roadside and bomb disposal units working on the moor. We were hard pushed to call this place beautiful; it was special but baron and repetitive.

The road twisted up to the top of a low range of hills. Family together, we looked across San Carlos Water. It was a short walk to the war cemetery and the grave of Colonel Jones, names of the fallen all about us.

More graves were in the ocean before us. Warship *HMS Clyde* came into the Sound, anchored and saluted the lost comrades of the water.

As we stood there in the soft grass, we could never have imagined the relationship we were to have with *HMS Clyde*.

We took time out for a tour of the small museum, weapons to handle first-hand — FM automatics of the Argentineans and self-loading rifles of the British. A Rapier missile frame lay in the middle of the floor. We enjoyed a cup of tea with the locals, then the battle plan lay before us. Helicopters lost at sea and troops ashore and on foot. The Marines pushed north towards Douglas Settlement. We followed the paratrooper regiment south, the target to knock out the airfield at Goose Green. In time of the war, the roads we travelled had not existed, instead it was boys with full packs TABing and yomping the grass land. Many tracks twisted down the Puzzle Gates.

Lieutenant-Colonel Jones VC

The edge of Stanley in the footsteps of Five Brigade

Local story was — Fire fight 250 rounds at Burnside Cottage, cease fire its locals, the dogs dead! We too took to foot and walked to the ground where Colonel Jones was shot and died. With machine gun holes all above the ridge, it was a wild killing zone and a wall of lead. Brave boys, a single paratrooper broke the line when a rocket grenade hit on the officers' trench.

A short distance over the hill and the small town of Goose Green lay before us. The town hall was the prison camp for the island residents of the time. Twenty-nine days of captivity. We retreated for stories of POWs and a look at the airfield, level grass for the cannon-shod Pukarar twin-engine fighters out of the Falklands sky.

It was homeward bound in the footsteps of Five Brigade Paratroopers as we made east for Stanley. The British boys walked by night in the cover of darkness. Each evening, they made for the hills that stood tall in the distance and marked the edge of Stanley.

We took a moment to take in the White House. The Paras found a phone line here. A local girl called Michelle took the call, and the troops advanced. Fitzroy and Bluff Cove, Welsh Guards and the Scots Guards were right of our flank. We made a push for Stanley under the shadow of engagements at Mount Kent, Challenger and Tumbledown.

War, war and more war tourism.

Weekends between school were broken by a garden party with the governor.

We stepped into the zone of the Governor's House and enjoyed the tea party. England was all about us, including croquet on the lawn. There was a Canadian piper by the veranda, flautists in the drawing room, and a marquee for teas, scones and cakes. Morgause and Caitland went with the Girl Guides for washing up duties, the bash raising funds to fly Falkland girls to Guide Centenary Camp in the UK.

A tour of the house with the governor found an Argentinean bullet hole in the wall, neatly painted with emulsion but left for posterity. Then it was up the narrow carpeted stairs to the snooker room. Days of old, cigars and gentlemen, darts and cues, and full-size Welsh slate green baize table. The underside was a joy to find. Autographed and signed by royalty and heroes, there were chalk signatures of princes and world dignitaries all about the underside.

A First Sight of South Georgia

It was in the far wing of the governor's residence where our future lay. Here we found the offices of the government of South Georgia, a sepia photo of the boss on the wall. Shackleton himself looked us in the eyes and beckoned us for adventure in the ice. At that very moment, our hearts were set.

Another weekend of excitement, and Friday, the last day of term, arrived in a flash. Destroyer *HMS York* slipped away in the weather, the local reception aboard ship cancelled. There goes the new frock — dash!

No new frock, but a fabulous T-shirt! Falklands Community School T-shirts were presented to us. It was house colours of red for Fitzroy given to Morgause, and blue for Ross given to Caitland. Girls together, they toured chums for signatures to remember their time in this faraway English school. Mrs Oliver, the maths teacher, added cream to the cake as she sorted a mighty folder of homework.

The weekend DVD for the end of school was courtesy of the South Georgia team at Government House. It was stories of Shackleton in the South Atlantic ice, the boss bringing his boys home from certain death.

That sepia photo on the governor's wall and the DVD was the key to the door of motivation that signposted that our next island would be South Georgia.

But the month of March was first. With school finished, we had a week of focus on all things tourism.

We began the fun with a bang. Mining expert Guy had a spectacular display lined up for us on

The ground burped before the bang!

Marzipan heaven

Surf Beach. His teams were methodically clearing anti-personnel material left from the 1982 conflict. They stored it up weekly and destroyed it with a show to match Guy Fawkes.

The ground burped as the sound of the Italian-made mines rattled our eardrums in their last stand before obliteration.

Saturday saw a horticultural show in the Christ Church Cathedral parish hall. Homemade cakes, best-decorated vegetable by a youngster, flower arrangements for the ladies, and jams in clear jars without air bubbles. Six of everything and keep them consistent. The sight of a Blighty British Battenberg was a magical image in our eyes. Carl was so overwhelmed that the Battenberg owner, Anna King, asked her daughter to hand it over for tea.

The local news was printed in a wooden hut, and the newspaper title was *Penguin News*. The only way to keep up with island stories was to go to the hut and get a copy.

World news arrived. An earthquake had hit Santiago in Chile; it was way up the eight scale. Clouds of airborne dust had closed air routes. The catastrophe had focused on Concepcion, and word was that Talcahuano, where we had so many friends, had been flattened. Ships had even been swept onto the land. There was no news on Robinson Crusoe Island, but belief was nothing remained of the village where our closest chums from the early summer lived. Natural disasters on the news had never touched our lives with such locality.

Earthquakes and dust clouds, flights all about the southern hemisphere were grounded.

A new chum, FIGAS consultant Nigel Allfrey, who was stranded without flights, offered his services. We loaded up his hired four-by-four and headed east. We went

The southernmost newspaper office in the world

CHAPTER ELEVEN: *Falklands Fun*

Falklands minefields

around the bay for photos of the *Lady Liz* shipwreck, a mighty red rusting wreck of a square rig steam sailor.

Wind and waves were a perfect combination for Barnes Wallace practice. Nigel took our minds from the disaster to stone skimming etiquette. We balanced rock pebbles and ocean. This was almost the same spot Brunel's *SS Great Britain* was rescued and refloated for its museum trip to Bristol. Back to the four-by-four, and we moved over the metal bridge of the Canache Lake beside the minefields.

From here it was into the land of the moon. This was as far as you could get on East Falkland. It was white sand and sharp white rugged rock on either side of a twisting track into the beautiful dunes.

We found a car park, portable toilets and a tea shop in an old bus. There was also a boardwalk path without timbers. Rope barriers of hemp directed us through the dunes. Magellan penguins lined the edge of the ocean water to the white sand beach. They were too little to trigger the Italian mines laid by the Argentineans. It was a treat of burrows and tuxedos to see the little fellows waddle about their day. We just had to join in. Nigel gave marks for best waddle as Morgause and Caitland skipped and shuffled down the path to lay an egg for a score over nine! We marched upward to the headland as the rocks grew bigger to defend the sea and saw a fine naval gun on the top.

World earthquakes, Argentinean threats and oil research could not have been more distant.

We spent the afternoon in Stanley for a look at the museum with the history of the Falklands — English land, French settlements etc. Then Spanish arrivals and the treaty of the River Plate, including warships of the First World War. Our trip had followed so much of this history. Repairs for Ajax after the Graf Spee engagements back at Uruguay in the Second World War. The trade wind routes of our own journey had intertwined so much history with our journey down here.

Outside the museum stood a refuge cabin. A donation from Antarctica, we were given the key to have a look at the bunk beds. Tinned food supplies, woolly jumpers and blankets, magazines and maps adorned the shelves. Life saving in the snow of the south and all left as it was from another era. A harpoon gun in the gardens sat by an Argentinean armoured car in the shadow of wooden whaling boats from South Georgia. The governor's uniform was kept pressed and pristine in a glass cabinet alongside a photograph of former Prime Minister Margaret Thatcher — autographed of course.

It was time for a trip to the lighthouse at Pembroke Point. Like the Antarctic cabin, you could get the key. We teamed up with schoolteacher Kathryn and GP Dr Rowlands for a joint trip. The key was a monster, more of a winch handle than a Yale. Troops together and four-by-four parked at the end of the road, we marched across the mountains of the moon. A wide-open seascape of white rock and soft sand held down the wiry grass. Then we saw the black and white stripes of a cast-iron tower. Morgause set about the lock for entry through the mighty door. *Clang, bang clunk*, and she opened wide. All together, we entered into the cylinder of wind defence. Sensible stairs to the first floor for a breath, then two floors so steep they have handholds in the vertical ladder staircase.

We could climb right inside the mirror Fresnal reflectors for a fun face in the photo.

The lamps were long gone. We had a wicked view of Pembroke rocks,

Looking into the light

CHAPTER ELEVEN: Falklands Fun

Wolf Rock to the south and how narrow the run into the bay looked from our light tower in the sky.

Beyond lay a memorial to the 1982 Falklands War. A giant four-bladed marine propeller commemorated the *Atlantic Conveyor*. With the names of those hands lost, the ship lay 90 miles and 60 degrees off the monument. Actions of the war were dictated by the loss of its cargo of helicopters, and this was the action that drove TABing and yomping into a whole new realm of courage.

Back to business and a permit for South Georgia.

It was time for a formal visit to the South Georgia office to check out that Shackleton material. We returned to his photo on the wall, as this was once his office in Government House. The deed was soon done, and we signed the application for a visitors' permit to the Island of South Georgia.

On return to the quayside, we had a new neighbour, the French sailing sloop *Hinayana*. Like us, they were also bound for South Georgia and Cape Town. Fifteen thousand miles of world tour on *Hollinsclough*, and South Georgia ahead as sure as the wind blew east.

But there was still time for tourism.

First we were set for a very special Friday — a date at Falklands military base, MPA. Schoolteacher Pamela sorted us transport, and what a treat she had delivered us too. Beyond the shadows of Mount Challenger and Tumbeldown lay globe-shaped antenna pods all about. The buildings were bristling with aerials and hangars dug into manmade bunkers of defence that escaped our view.

Security cleared and photos on the computer, we found ourselves inside. Flight Lieutenant John Davies, "JD", greeted us. Seats in his Landrover awaited for a tour onto the tarmac. This was no ordinary tarmac. An FOD (foreign object damage) check was made on every tyre for gravel. This is no place to leave a flying stone for a jet engine. Airstrip level, FOD clear, landing lights and marker posts port and starboard, we took a drive down the runway.

Oilrig exploration 100 miles north of the Falklands found us in the middle of a battlefield. This was a state of alert; we were in downtown Top Gun territory. JD cleared us through inner security. Razor wire fencing behind us, we stepped into the QRA zone (quick response action).

Four hangars lay before us, and each hanger housed an Eagle call sign of 1435 Flight. JD smiled, "They have names — Hope, Charity, Faith and Desperation." Family together, we walked onto the hallowed ground of Charity — call sign Eagle One. Her tail fin carried the Maltese cross overlaid with the eagle. She sat proudly in soft matt grey livery, her delta wings hunched up with fuel pods to cast shadows

Her name was Charity.

of splendour across the hangar floor in the afternoon sunshine.

Charity had her small control wings lying softly downwards at the nose cone end of the bird. Yes, she was a full-blown war ready Eurofighter. That's a Typhoon to you and I. We had Googled the info and checked the spotter book, but nothing quite prepares you for that nose cone of wing flaps that turns instability into daring for a dog-fighting machine of destruction that can knock anything out of the sky. Trimmed for direct air-to-air combat, an all-missile payload defined her as Top Gun southern hemisphere. Challengers should not proceed, not pass go and report directly to jail.

Our Monopoly board was loaded with technology to die for. JD introduced us to the G-suite clothing, immersion survival togs, and air and helmet systems. We took a tour of the missiles, long-range radar and short-range heat seeking — don't mention the cannons when they get up close for the whites of your eyes.

We walked nose cone to stern, composites and titanium, laser guidance detection systems, anti-missile heat flare tubes and more wing racks than a rib of steak for those missiles. We had a look at the parachute brake and the emergency hook before we found the twin Rolls Royce afterburners. Then it was up the port side to see the starting system and take a look at the ground engineer comms hook up.

Stairway to the stars, Morgause and Caitland walked up the loading ramp. We couldn't believe the moment JD offered them the privilege of sitting in the Bang Seat — auto throttle in the left hand, electronic stick in the right, heads-up display in front of their eyes. Memories are made of this. Charity was turned on, her systems were up, cocked and ready to go.

A green light filled the cockpit as it flooded from the displays. Chart plotter in centre screen, radar to port, altitude and systems to starboard. Flexing fingers on the throttle switches, the cursor ran about the screens as the zoom locked into Falklands Sound and the way points for refuelling. Panic data was low on the starboard side, and failsafe gauges flipped out top right. Totally epic, this was the real thing.

Squadron Leader Richard Wells arrived. He refocused reality with an invitation to the scramble room. Two Eagle call signs were always on immediate standby. The ceiling of the room was draped in a huge Union Jack. Mugs of steamy hot tea and stories of the Typhoon trip from Blighty with hourly refuels in the sky leaves only one question: "Where do you go to the toilet?" Memories of this will last forever.

Back to the future for a change of time

Back to earth from the privilege of our warplane adventure, in the fifth and last week in the Falklands, there was no respite from the excitement. The church clock needed help, the *Pharos* patrol ship was in, and a marathon was about to take place before we would leave this land of England in the South Atlantic.

It was time to go up the tower with the vicar's climbers. Carl helped Major Peter Biggs and his son Kyle with the church clock; climbers being supported with boat ropes and blocks was fabulous fun. The dials of the time machine had to come down for maintenance. By mast climbing, it was a walk in the park — shackles, rope and pulleys. Unlike a mast, the spire stood stationary in the wind. A timeless photo of clock tower heroes was fit for the mantelpiece.

The fisheries ship *Pharos* arrived at the end of the FIC jetty we had made our home, as it had a few days of engineering work scheduled. Skipper Chris and his team welcomed us aboard. A tour of the bridge followed. Today Caitland took the skippers chair but we were on the ropes.

We had never seen a diesel-electric power train before, mighty yellow diesels and enormous magnets of electric motors. There were enough megawatts for all of Stanley. The *Pharos'* key tasks were to do monthly shuttles to South Georgia, patrol the fishing vessels under licence and deliver scientists and supplies to South Georgia.

Pharos bridge under English girls' control

With ourselves and French yacht *Hinayana* all together, we were forming quite a regatta for the South Georgia run.

There was one last full-on shopping tour. The quayside was a supermarket from England.

We then had a Saturday evening pasta party for the Sunday marathon.

Sunday March 14th, Mother's Day for the southern-most certified marathon in the world. The Stanley horticultural show provided the cakes for the mums' breakfast, but it turned out to be a day for the dads. Best frocks were team T-shirts, a capital D for Dad, and every office print stamp in town had been collected to adorn the decoration. All Stanley was running with the dads' team — from the petrol station to the post office.

Our first gangplank to the Pharos family

CHAPTER ELEVEN: Falklands Fun

Tamsin's dad, Mike Mcleod, was on the first leg, French papa Jean-Yves, 10,000m champion on the tough Sapper Hill, had the second section. Lizzie's dad, Phil Reed, was on the downhill dash, and Hollinsclough dad, Carl, took the finish run.

A proper dads' team. "FAT fat fathers', surely not. We entered as the

A little English shopping for the tinned stuff

Hollinsclough Harriers. DADs! *D*ashing, *A*thletic and *D*aring, Dads! That was the children's chant. Cheer leaders extraordinaire, it was the children, heroes of the race and great motivation, who set the day. There was a Sunday morning team meeting at nine sharp with a shot of Caribbean rum to set the dehydration in place. Dads and youngsters grouped for the glory walk down the waterfront to sign in at the town hall. Strutting and stamping, sweat and excitement, the moment of truth was upon us.

Bang went the gun, and the DADs were away, arm in arm for the first part of the first leg, flags flying and yacht colours adorning the sky. The dads all faced the wind and dug in for punishment. Mike had just finished his twelve-hour work shift but delivered the goods on a stomach full of porridge, making a waterfront run and then up to the town hill. Sarah Crofts did the taxi work for the back-up team. No training and no notice, he still bagged the first 10k inside 90 minutes for a credible start. French Champion Jean-Yves took Sapper Hill by storm and moved Hollinsclough Harriers up the field with a magnificent push into the wind fit for Seb Coe and Steve Ovett. Drink stops manned by locals offered vital refreshment, but none was better than Mrs Cooper and her bacon butties and Margaret's hot coco. A VIP runner from Kenya set to take first withdrew in the freezing wind. Ryan Elstow's taxi shuttled Phil to take the handover just short of the Sapper Hill summit and knocked out the top for a downhill sprint. Suffering from leg cramp with the decent, Carl rallied him half a mile and then the youngsters took over to bring a recovery and deliver him to the final leg. Carl closed the airport, came about into sleet, placed his diving goggles into action and took a few late placings to push the flat lands for

Fat fathers in the southernmost marathon on Earth

the FIPAS ship dock. Flags in hand, it was a full team effort from the Jetty Centre as every child of the dads made the Waterfront sprint. Landrovers on the road flashed their lights and hooted horns.

The dads team were no athletes, and in the face of adversity, they must have been the heaviest team in the race. Jean-Yves joined Carl to push him hard for the west turn beyond the museum. Legs were flying and by the *Penguin News* building, both Phil and Mike joined arms. The four dads together carried each other down the last mile.

Mars Bars, hot chocolate and bourbon biscuits provided the celebration. It had taken us a little over five hours, with an average temperature of five degrees and average wind strength of force seven for the southern most certified marathon, the race exposed to the Antarctic winds and insulated with fat fathers.

The Army Ghurkhas took the first team prizes, the RAF rock ape was the over all winner, and the Navy commodore saluted all the runners and their efforts as he handed out the prize money.

Marathon done, but we had to leave a friend.

Our days numbered, we had a friend to leave on the Falklands. Time for a family dinner with the vicar, and we met George the magic cat. Our task was to hand over Rene the reindeer. Rene was our Christmas light decoration reindeer. We had found him in Chile, and like our beloved toaster, he was 220 volts. A marvel of tiny lights and twisted steel, he was a fragile chap. The Southern Ocean was no place for him. We handed him over to the vicar, and Rene was at his new home in the vicarage. We dreamed of him safely twinkling in the decorations of next Christmas. We celebrated with chocolate fruit pudding, homemade to perfection.

Returning home to *Hollinsclough*, we slept sound with tummies full, ready to leave this English island in the morning.

Chapter Twelve

Falklands, bound South Georgia

Summer sunshine in Cumberland bay, South Georgia

"*Ahead lay a 900 mile leg of gale eight winds...*"

"*Nothing quite prepared us for the enormity or beauty of these beasts.*"

"*Captain James Cook '..the wild rocks raised their lofty summits till they were lost in the clouds and the valleys laid buried in everlasting snow. Not a tree or shrub was to be seen'.*"

Goodbye Falklands. We sailed on St Patricks Day, 105 degrees bound for South Georgia on the South Atlantic run. We were south, deep south, with Antarctic convergence currents and water so cold that the toilets freeze.

Sue organised the FIC Speedwell launch boat, it gave us a tug off the FIC jetty into the wind as French Hinayana rib sorted our ropes.

We cleared the mighty red bow of the *Pharos* fisheries ship and waved goodbye to our French chums aboard *Hinayana*. We made a quick run west up the harbour for a perfectly timed twelve-o'clock wave to school chums as they flooded out of the school front doors for the first lunch break of their new term.

We exited Port William Sound from the Narrows at one p.m. local with a logged position of 51.40.00S 57.49.00W.

Ahead lay a 900-mile leg of gale eight dash to South Georgia, but then that's the South Atlantic for you.

Ten minutes past two, JD came by for a wave in Eagle One; the Typhoon Eurofighter was so low he was touching the waves. He made a tight circle to

Speedwell pulls us into wind from the FIC jetty

return for a wave, rolling his wings over the back deck. We had seen nothing yet. The radar was on to give him some target practice. He screamed in direct to the bow and went vertical over the mast. The pair of Rolls Royce motors spat flame as the afterburners rocked our stomachs with a sonic boom. We were stacking up life memories, and that was one for the mantelpiece.

A big wind evening howled past force eight and topped out at 49 knots of wind for a fast first night sail that would notch up record speeds growing by the day. Jib on the pole, cutter full and a little main for steerage. Our memories of Uruguay were strong, Jim Kilroy of *Kiloha* fame had cautioned us on the main: "No mater how high the wind, be sure to have a little force into the back of the mast for best steerage."

A clear evening of wild starry sky rolled into a huge sunrise. Albatross and dolphins were around, and it was Pot Noodles for a late breakfast. The back-up generator set in the rope locker worked a treat to boil the electric kettle for steamy mugs of tea and coffee.

The first 24 hours delivered 170 nautical miles run And put us at 51.21.50S 53.24.00W.

Pushing west, 105 degrees and running six knots on the wind for the daylight, we celebrated with hot stew for tea — best mince from West Store Stanley on the Falklands. The wind was calming a little, but walls of water like the rolling hills of the Peak District were building for sunset.

On occasion, *Hollinsclough* heaved up out of the ocean for wild views on the top of the crest of a rogue. A monster, grey ocean crashing in all directions surrounded us. Then back down below, most of the view without horizon, tunnel vision of a building storm circling from deck to mast top.

At bedtime the wind was steady in the darkness, and it was time for a little help from the Perkins iron sail about midnight. We were quick to fire up the motor and get the bonus of topping up the batteries with that 60-amp alternator. We had very good radar images, but it was getting cold. Fully togged in sailing kit over extra tracksuit trousers, still we wrapped in a full-blown quilt to sit inside the canopy for darkness watch on the computers.

We did six-hour shifts — six to midnight, midnight to six — watching space invaders on the radar screens. "All things bright and beautiful" was played on the teeth in the chatter of the chill.

We had wild-red morning sunshine for Friday. There was no fish at Woodbines Fish and Chip Café in Stanley today. The smell of the burger van at the FIC visitors' jetty was far away but still teased. Outside of the range of the Typhoon fighters, we were all alone.

Eleven local at 52.43.00S 49.33.00W, we had 100 degrees on the dial at seven knots. We were 300 miles out and 480 short of Grytviken, South Georgia. There was nothing on the dial in any direction anywhere. About halfway down the Falklands to South Georgia leg, in the absence of any shipping, we began to realise just how alone we were.

Even the ocean seemed to carry a postage stamp marked far away. The swell and current was in there somewhere, and the run of water was lumpy, waves all confused, but the ride was still sensible. We had lots of sail still out to harness nature and stabilise the yacht. Sea sickness behind us, sleep deprivation under control, there was no fish and chips, but we had naturalised to the environment.

Friday evening was a long, dark sail on steady 25-plus wind. The target was west, but we were running as much as 120 degrees for stability and comfort. A little too far south, the Antarctic convergence zone of ice-cold southern water chilled the evening. The radar scanner was set on three miles for iceberg watch.

Around midnight Friday March 20th, we recorded 15,000 miles on the distance log. The dark evening watch routine was one of numbers, counting, writing and calculating.

We were happy to see the six a.m. Saturday morning sunrise — grey but with good visibility. We had the motor on for some of that elusive battery charging. We tacked to port at 53 and a half south and began hunting back some of that west. It was a smoother ride but very confused waters. Grey and cold, a little drizzle, but how we cheered up with biscuits and coffee.

Saturday at noon local Falklands Time, we were at 53.22.50S 45.00.50W. We were 110 miles west of Shag Rocks, looking to run north of the first outpost of South Georgia and daring to begin arrival predictions for log, time and travel. Key to this end of the run was collision and ice.

The afternoon winds steady, we contemplated the night watch, all too aware that iceberg country was coming. A Saturday evening surprise took us off guard on the AIS computer. There was a vessel ahead. We hadn't seen a ship on the whole run up until now.

At 18:00 Falklands local time, the AIS computer was the first piece of kit to spot the 350-foot passenger cruiser *Prince Albert II*. He was out of South Georgia heading to the Falklands in reverse to us. As range closed, we chatted on the VHF, "All is well." That was a message to warm our hearts.

The best news was he had seen no ice since the edge of the islands. We left him battling the wind that was running well for us. We had great warmth to our minds after that advice of no ice tonight.

We past north of Shag Rocks a little after midnight, no benefit of a view in the black darkness. We had layered another tracksuit trouser over the first this evening as we stiffened in the cold.

The seas for sunrise were still confused. A bumpy breakfast, we were sailing just on the cutter with strong winds. Our daily mileage was breaking all records, but the average wind had been gale force eight for every one of the days. The familiarity of the environment drove attention to detail. Everything was repeated to the hour, on the hour and every hour, miles counted in the log to the very last detail.

Six a.m. Falklands local, our position was 53.13.50S 41.25.00W bound 90 degrees at 6.5 knots.

It was a long morning of bumpy seas, and Tracey took five washings in one hour of her early watch. That's the whole cockpit flooded from rogue ones coming in off the port stern quarter. Thank God for a narrow stern; it only pushed us out of autopilot once!

Wet thermals in ice-water cockpit watch, there is nothing better for character building as the teeth rattle out those lines of "All things bright and beautiful."

A long morning, long day, force eight all the way. The Grib forecast showed a break late evening, and the miles rolled up the log as fast as the cold dampness blurred the ink.

Joy of joys! In the early afternoon the pressure stabilised and some sunshine drove through the grey clouds. Gosh, it made a difference to the temperature. We all smiled at each other with steamy breath.

We dreamed of putting the heating on but were too cautious to damage it. We were not using boat central heating, very aware of damage to the oil burners with the boat being kicked about so much. We were sure to need the heating in South Georgia. Everyone took camp in the back bedroom, as at the front of the boat it was freezing cold. The stern was warmer with the heat from the Perkins. Every run for the batteries aided the warmth.

Afternoon sunshine in the deep south, a heat wave of five degrees was on the gauge. The portable generator fired with the first pull of the cord. The house generator set hated the boat at big lean angles; it would suck air on the sea water cooling pipe and overheat before we boiled a kettle. The temporary generator set in the locker was an air-cooled machine. Its reliability was a dream, and it ran at any angle, even the five-degree heat wave wasn't going to overheat it.

With easy electricity on hand, we microwaved a steamy tin of Heinz Spag Bol. Joy of joys, energy restoring. Cake and coffee followed. We decided to run Perkins for a few hours to help push more warmth into the boat. Hot water bottles from the generator run were all bliss.

Life was getting better all the time, but the skeleton in the cupboard, the darkness in our minds, was the absence of vision as nightfall delivered our most likely iceberg encounters.

Time for decisions. The gale force weather out of the Falklands had given us fabulous mileage, but it had worn us down. We had had five days of damp cold and record-speed sailing.

We were not keen to heave to and sleep an evening stationary at sea in this confused water. We were well north of Shag Rocks, so the plan of action was to just run slow and stay well to the north of the coast of South Georgia. That would give us the chance of a hard daylight run down the coast to be in Grytviken by next evening's darkness watch. We would save the danger of another night beyond that for the last run.

Cruise ship *Prince Albert II* reporting no sea ice spotted pressed our decision. All this wind must have driven any ice farther away, so by our calculation, the way should be clear and the ice driven back in the hard wind.

CHAPTER TWELVE: Falklands, Bound South Georgia

As that Sunday evening darkness fell, we drew back to a short cutter and very little main for steerage. We set the radar scanner at one and half miles and switched intermittently close up for three-quarter-mile detail. This was real computer game sailing.

It was hard to slow the boat down below two and a half knots. There must be some current; the water was clearly pushed hard by the wind drive, not tide. We were emailing Roy back in Blighty to control any rescue events in the case of disaster. It would be a long evening for all of us.

It was a classic, black-ink, no-vision night as the yacht rolled about, running too slow, and every wave crash sounded louder than ever.

Around midnight for watch change, I had a look at the twelve-mile scan and picked up two groups of strong targets. That meant big icebergs on those range settings. It was frighteningly cold for the night watch, but it was easy to concentrate with so much emphasis on the likelihood of ice. I was now running two pairs of gloves, as any steel on the canopy rail or winch sets stuck to the skin with the cold.

Three a.m. for a dance in the dark, there was something on the three-quarter -mile scan. Repetitive shadows in the scatter, ten degrees on the rudder, then 20 to be sure. Scatter or solid, we will never be sure, but it was all caution to come about and move the line.

It was a balance between paranoia and radar scatter. We shortened the cutter to near nothing but were still unable to get below two knots. We even considered a little reverse on the motor, but my reality was that the movement was relative to everything around us, no ice was going to be coming the other way.

It was the most odd feeling to be wishing down ground speed, trying to get rid of distance in a life where we had focused on fighting to gain forward motion which nature had so often denied us.

Nerves frayed, the soft light dawn couldn't come soon enough. The enormous relief of penumbra came about five a.m. The email to Roy said, *"Get some sleep, we are on daylight watch. We are still alive with no Titanic berg crash. Fifty-three south in Antarctic convergence ocean, South Atlantic. 53.29.15S 38.10.10W."*

The plan to *not* heave to was looking good, and one long daylight watch to win the prize was within our grasp.

Monday morning, and what a perfect day to make that dash down the coast of South Georgia. With all the energy of a five a.m. sunrise, it was easy to forget that most of the world would still be in bed.

With daylight rising, the water was relatively flat for down here. After so many days of gale force eight, it must have been a start of week thing. Wind falling with

Mogberg sighted, and it's cold enough for the balaclavas.

pressure softly rising, gosh, this seemed like good weather just when we want that last push in! Time for the motor and all speed.

Downstairs in the yacht was like the Tom Hanks *Apollo Thirteen* movie when he fires up the spaceship — cold, damp and just frighteningly freezing.

Icy water on the switchgear, Polly Perkins the metal motor sail started on the second turn — well done her. Still cautious, we gave her five minutes to air before throttling. Sunrise proper was about five forty-five am and visibility was maximum.

By six a.m., there was almost no wind. The cutter was set out tight, lots of main on boom to an arrestor for a seven-knot start on the power of the propeller. The fifteen-knot record speed run from a few gale days ago was long gone. Softer days we had rarely seen anywhere in the world!

Six-hour watch shifts for this last day were lost; we were unable to gain the sleep patterns back with the excitement of the job in hand and committed to a stay-awake pattern for Grytviken.

More groups of bergs began to appear on the long-range radar. Nothing quite prepared us for the enormity or beauty of the beasts in daylight. They were like a small Channel Island — Alderney or Sark. Totally enormous, you could have lived on them. The first one got the name "Mogberg", sighted by Morgause. "Kitty Berg" for Caitland followed, something looking like Dougal from Magic Roundabout, it was the size of a small town.

Wild white sculptures carved by ocean, land and glacier, they were very much whiter than we expected — no blueness of the Patagonian glaciers.

The whole place was so cold, and these bergs were so enormous they even had their own cloud formations. Small ice growlers following, bobbing haphazardly about the water, waves and rip tides licking the shape away. We watched a variety of debris tailing the berg lines. Some debris was so small it would dance in the waves and only show itself as you looked down from the deck rail. How could you ever avoid that in the darkness?

CHAPTER TWELVE: Falklands, Bound South Georgia

Having seen how small some of this debris was, it was time for five-minute eyeball watch. Five minutes head in the wind over the canopy was as long as was bearable in the freezing air. We took turns not to have a single gap in the hours that followed. Our fingers were stuck to the canopy rail, as the steel was well below freezing. Gloves were a necessity as they tore on the cold metal.

Breaks in the surface of the frozen water stopped your heart instantly. It was not all danger, land was near as

The first sight of land may have been Cape Buller.

we saw the cheeky grin of fur seals leaping about their playground. Torpedo streaks through the clear sea popped up as the hourglass dolphins shared our bow wave. Pointy beaks followed, and an unusual bright orange glow broke the snow-capped waves and disappeared. These beaks were kings, not Prince Charles but king penguins. Without any technology, it was clear we were nearing our destination.

Caitland was first to spot land; it may have been Cape Buller.

We didn't call her "eyeballs" for nothing.

More and more fur seals larked around the bow, playing with our arrival as the large white peaks began to grow in the distant cloud base. Mileage targets and speed over ground were good. We set the autopilot tracking line direct for Cumberland Bay, and the time estimates gave us a sunset arrival window. The track line and projections left nothing to spare as we calculated the sunset for eyeball vision on the last entry to the coast.

Captain James Cook: "*...the wild rocks raised their lofty summits till they were lost in the clouds and the valleys laid buried in everlasting snow. Not a tree or shrub was to be seen.*"

His words could not be bettered.

This place of Southern Ocean Antarctic convergence zone was a wild wonderful panorama of nature at its finest.

The darkest night of iceberg alley was over, and the brightest day of our lives lay before our eyes.

The mountain-scape of South Georgia was the finest panorama we had ever viewed.

Patagonia's Beagle Chanel, even the gallery of glaciers was nothing on this. There were mountains of monster size, as many as you could count, jagged and as sharp as the teeth of an ageing shark, an endless glory of land against sea. Family together, we peered out of layers of balaclavas to take in the picture as nature fought to freeze our eyeballs in the wind.

Years later, Morgause would paint an art picture at school. Asked for mountains, she replicated the scissor-like monsters in their multiples. The teacher said, "Don't be silly, nothing such as this exists."

Below decks *Hollinsclough* was warming with the motor running. All sail was out, and as the wind fell to nothing, we had become so brave as put the heating on. It fired up well and warmed the dampness of cold. What a terrible stink followed as the damp clothes hanging on every rail peg and shelf aired.

The tracking line took the autopilot progressively closer to the edge of the mountains. Confused by the stature of enormity, our eyes found it difficult to judge the range. The mountains just rolled up further into the sky without the coast getting nearer. The shadows of the end of day began to fight with their tops, and the twilight darkness grew frighteningly faster than the clock could chase the shadows.

Cumberland Bay opened up bang to map as we swung starboard into the teeth of this magical landscape. The VHF on the radio was as crystal clear and Blighty English as any marina on the south coast of England.

Icebergs of the South Pole lay aground all around the bay. Grounded on the edge of this world, guarding it like soldiers on a battlefront.

The darkening sky twisted into piles of lenticular cloud deflecting off the mountains far above, and with the last breath of light, we dropped all sail and motored into King Edward Point jetty under the shadow of Shackleton's memorial cross. The shadow ran long in the very last of the day's sunlight.

King penguins and fur seals were all about. A mighty elephant seal moulting in the tussock grass stank even more than our damp kit, but what a wonderful smell it was to be tied safe ashore in this land of nature's most deadly ice.

King Edward Point, Southern Ocean, South Georgia – 54.15.00S 36.26.60W

Government officers Kirin and Robert sorted our paperwork. We were tied tight to the concrete jetty for the evening in this faraway splendour. It was time for bed and the heating was on.

CHAPTER TWELVE: Falklands, Bound South Georgia

Ashore South Georgia, we were met at King Edward Point by two lines of porta-cabins housing the scientists of BAS, the British Antarctic Survey. They would become great friends, but meanwhile, visiting yachts tie up across the bay.

A marina home at Grytviken Whaling Station

Not an hour beyond breakfast, and we were quayside at Grytviken, the old whaling station. Not a thing had changed from the day they closed it down in the sixties. The only ones absent were the crusty old whalers of Shackleton's day. Large buildings with tin roofs rattled in the icy wind; they all shone golden to greet us to this macabre death zone of the whales. Autumn gold warmed to the colour of rust that carpeted the age of the place from its towering whale-oil cylinders to the old generator engines stood on their shore-side concrete. There was a modern museum building with a house for the curator, Ainsley, and at the foot of the mountains, a white church with tall steeple marked the path to Maiviken Bay beyond.

On the quayside, we had a new neighbour. David and his circumnavigating yellow explorer motor yacht *Polar Bound* lay ahead of our bow.

The golden colours of autumn sunshine marked a glorious time in South Georgia to befriend the scientists, live with nature and be part of the history.

Our first day was a homage to Hope Point, paying respects to the boss. It was time to meet the scary seals on the Snowdonia-like path. We took easy steps through King Edward Point BAS science town, then between the teeth of the monsters. The fur seals in the Pacific were friendly, but these were wild animals. It appeared the seals of South Georgia had memories of the whaling station and were ferocious. A sturdy confident step and a big walking stick in hand was a necessary defence against these chaps as we placed our steps with all care amongst the not-so-friendly creatures of the island.

CHAPTER TWELVE: Falklands, Bound South Georgia

Our toils were celebrated with a salute to the cross of Shackleton. Our respects paid with a prayer for the boss, who brought home his boys from the danger of the ice.

Afternoons that followed were island explorations. With each day, we became more confident with the fur seals. Notice given was the best action; surprise one of the little monsters, and they would delight in a piece of leg.

We made for Penguin River. Fur seals by the hundred, king penguins, smelly elephant seals, beach masters of monster proportions with teeth fit for vampires. Collecting photos as we travelled, each and every one a mantelpiece nature prize winner right before our eyes.

There were icebergs, mountains, whale bones and ocean to frame every photo.

Wherever you walked on the shores, there were whale bones all around, a macabre eerie memory of a past of carnage unacceptable to our modern world.

Moving history forward to the Falklands, we marched up the hillside to find a helicopter crash site. Perfectly preserved from the Argentinean attack, there were bullet holes around the untouched instruments. With no visitors, it was as perfect as a display in the helicopter museum back in Weston-super-Mare in the UK.

Back at Grytviken marina, it was positively busy. Single-hander Ron arrived on American yacht *Relentless*. The remote quay was packed full with three yachts. We were quick to march him confidently past the fur seals: "Watch for the whites of their eyes." Shackleton's grave was in the white cemetery of the whalers' cemetery, and all together we saluted the boss.

There was something about the Shackleton story of ice survival that brought home the true ethos of this iceberg mountain of an island.

Helicopter wreck from the Falklands War

Grave of the boss – Sir Ernest Shackleton

The Southernmost Girl Guide meeting in the world for the 100-year party

Letters home! On a visit to the post office at King Edward Point, we met Ruth, the wife of government officer Kirin. Ruth was the postmistress, but more importantly she was a former Brownie and Girl Guide, Cambridge Troop, Swallow Patrol. She introduced us to Susan Woodward, the doctor. She had graduated out of Newcastle and world travelled from African hospitals to Borneo for multi-continent medicine. Yes, sure enough, Sue was also a former Brownie. British Antarctic Survey boss Ali was a former Girl Guide too.

It was time for a blue mafia party, the southern-most Girl Guide meeting in the world.

With perfect timing, Girl Guiding UK World Thinking year for 100 years of celebration was upon us. This meant best Blighty neckers for an afternoon tea party. World Guide badges were honourably awarded to the southernmost Guide meeting in the world.

A giant grey ship dropped anchor in the bay. Two chaps were at the window! Ben and Ben, the navigation officer and the weapons officer of the Royal Fleet Auxiliary Ship *Wave Ruler*. The giant military tanker ship was visiting South Georgia for its very first time, preparing for the arrival and refuelling of Royal Navy Frigate *HMS York*. What a treat, tourists in this faraway place.

We said hello to the boys, talked of our own adventure and pointed them to the gravestone of the boss. Respects paid, Navigation Offi-

All aboard RFA Wave Ruler

CHAPTER TWELVE: *Falklands, Bound South Georgia*

cer Ben returned with an invitation to visit the mighty grey war ship tanker throwing its shadow across Cumberland Bay.

Family together, we togged up in survival suites to get aboard an all-weather grey launch. It sported a soft roof cover for southern sea spray and sped out past King Edward Point jetty. We donned a safety harness and lifejackets to climb the ships port side-boarding ladders.

We followed enclosed stairways from the boarding deck topside onto the floor of a small town. With a wide-open expanse, it was a full-blown spaceship in its own universe. Pipe work, steel cranes and gear were all about. We went forward for a look at the refuelling cranes.

Fuel forecourt for destroyer HMS York

The pipes for aviation gas, diesel fuel and oil sludge all had special marking colours. The ship was at work and a guest was arriving.

Our visitor was greeted with a stainless steel rope wire, and we awaited the berth of a needy ship. Navy and rifles, SA80 plastic pistol bullets fired the first line for attachment. The gas turbine rocket ship *HMS York* pulled into the forecourt for fuel.

The ships were side by side mid-ocean. We were privileged with the great honour of being part of this exercise. It was military Royal Navy to perfection, not one drop spilt. *York* powered up for a destroyer speed fly by.

Our tour of *Wave Ruler* followed. A structure large enough for most ship bridges was the crane command station for four fuelling crane pylons — Star Wars eat your heart out. All forward for a run up to the most forward raised deck. Six-ton anchors hung on a giant chain. Marker links adorned in red paint, length identified by links painted white. It was so similar to our own but on a scale beyond belief. Eight white links made eight shackles.

Aft please, and 650 feet astern for a full deck walk. Thirty-mill canons for defence, watertight hatches of mass proportions, then a helicopter deck the size of a football pitch — space for a Merlin was no problem. It was then indoors through the flight hangar — most truck garages on land are not this big.

Coats hung in the officers' mess, and a fantastic tour followed. We walked a posh corridor with photos and thank-you plaques down the wall of honour. A look at the fuel bunker bridge computer room, then the war room with radar and weapons systems, armour proof and red lighting.

She was old enough to respect such a privilege.

We then went upstairs to the bridge proper. You need binoculars to see the other side of *Wave Ruler*. They had mini-gun cannons that shoot 300 rounds a second for more defence.

The four-bar captain welcomed the girls like family. Morgause took the skippers seat. She was old enough to respect such a privilege that would become a life memory. "Don't get comfortable young lady."

Girls and officers on the bridge of RFA Wave Ruler

It was a very modern bridge, radar displays sat side by side for best definition. With a single propeller, single rudder, bow and stern thrusters, no tugs were needed thank you very much. The electric motor with diesel generators meant one and a half megawatts ticking over at anchor. It was like a power station in a small town.

CHAPTER TWELVE: *Falklands, Bound South Georgia*

A Blighty treat of all treats was in store. We were invited to the captains' table for chip shop Friday lunch. Tarragon Pan fried plaice fillets, tartar sauce and lemon wedges, chipped potatoes and mushy peas. It was Blighty Friday food at its very best, and all in this far faraway place of nature's greatest splendour.

A million thanks to Navigation Officer Ben for a fantastic Friday.

Back at Grytviken marina, a fourth yacht had arrived. *Wanderer III* was moored amongst the pontoons with Teece and Kiki.

HMS York lads in Grytviken with the girls

Then *HMS York*, completing a tour of the island, dropped her anchor in Cumberland Bay. The boys were soon ashore to say hello.

Rev Ralph, the Royal Navy padre, came ashore. Ash Sunday had gone, but he was bang on time for an Ash Monday service in the church of Grytviken.

The best grass was mowed for an afternoon football match — civilians and scientists against the Navy. Skipper Sunray of the *York* joined our side to help out for almost a draw as Government Officer Robert defended to the last in a mud battle where the field began to thaw with the excitement.

HMS York football team against Grytviken civilians

We made so many new friends at church and playing sport before the *York* raised her anchor. We dipped our ensign in salute, and she sailed for warmth, heading north.

With the exit of *York*, an old friend arrived in the bay. Fisheries vessel *Pharos* was in the harbour. She tied and chained to the KEP base concrete and dropped her boarding ladders. We had a full English breakfast for a feast fit for Blighty. Then it was a morning on the bridge with Chris and his team to compare notes on the gale-driven tides of the Southern Ocean that the *Pharos* made its own.

Fondly named the "Big Red Ship", she also delivered supplies to BAS. A barbeque party for the base locals was a family affair. Sizzling embers in the top cargo hold, and building snow whistled in the hatch edges where the smoke of the barbeque escaped. Burgers warmed the stomach as a roaring gale raged outside. The scientists faced a 100-yard walk home, whereas we had to circle the bay for Grytviken. The girls were family to the *Pharos* crew, so concerned for their walk back as the wind gauge roared past 60, Chris said for us to stay on the *Pharos* for the night. There is nothing like going home to your own bed. The *Pharos* boys took battle stations and shone the iceberg search light around the bay path. That rallied the fur seals into a fit of frenzy, but they were suppressed by our giant shadows. Reflections of

CHAPTER TWELVE: Falklands, Bound South Georgia

light danced into the mountain walls as we bounced around in the wind for our warm beds on *Hollinsclough*.

A snow-white carpet had taken our footsteps away by breakfast as another yacht arrived in the bay in the sunrise. The light of the sun here in South Georgia gave us more solar panel electricity than anywhere we had measured in the world.

April 1st for no fools' day. It was French yacht *Hinayana*, Jean-Yves and his family of four children arrived from their Falklands leg. Golly, the place was a busy marina in the Southern Ocean, but this was no place for a fat fathers' marathon run.

We organised a sunshine walk to Gull Lake. The icy water of the rich blue lake provided power to turn a modern turbine and give plentiful electricity to the scientists. Beyond that lay nature.

Good Friday, explorer cruise ship *Professor Mekavich* arrived for the day. In our bustling mini-marina, we had seven children for the Easter egg hunt. It was a true family affair of fun as adults laid eggs and clues in the whaling station. Ainsley in the museum finished the hunt with a word game from the clues. The answer was "British Antarctic Survey".

On Saturday April 3rd, we saw our the first winter iceberg in Grytviken. A white monster had sneaked all the way through Cumberland Bay to arrive ashore as a new neighbour. The season was at an end, and it was getting time to leave.

A mountain walk with Dr Susan to Maiviken on the far coast. The path beyond the church was easy, Susan then showed us the way over the broken shale of the mountain. It was bog lands of the Lake District and lunch in one of the survival huts. Then we went over one more hillside for Southern Ocean views. Penguins, albatross and seals added splendour to this faraway place. Nothing ever prepares you for the sound of nature in these bustling wildlife homes.

Easter Sunday, and we had chocolate Easter eggs — well done Stanley shops. The Ship *Ernest Shackleton* arrived, so our supply of Pimms, cognac and toilet rolls was replenished. Chief Engineer James Shaw showed us around. On all our tour of the world, we must have visited more ships in South Georgia than in any other port in the world.

It was as they left it.

It was that first iceberg aground in Grytviken marking April that showed us it was time to say goodbye. More mornings than most, we began to find any ropes left out had frozen solid to the winch, and on the colder mornings even frozen to the teak timber deck.

Snow had begun to engulf our world, the whaling station took on a new look as its rusty steel shed snow to contrast with the world around it.

Tuesday April 13th marked time to go. The plan of action was ten days to Tristan de Cunha, and then a northerly swing for Napoleon's exile island of St Helena. Twenty days at sea, 3,000 miles and all getting warmer. Ascension, Cape Verdi and maybe the Canaries for a September school run.

CHAPTER TWELVE: Falklands, Bound South Georgia

As the ocean snarled up with white-crested teeth, our Golden autumn memories warmed us. *Hollinsclough*, South Georgia, Southern Ocean, an oasis of sub-Antarctic Island, Albatross, fur seals and mighty elephant seals. There were penguins of every variety and even Norwegian reindeer.

We had made Tim and Pauline Carr's book "Antarctic Oasis" our autographed memory of chums on this faraway island. The book was busting full of signatures to carry those memories.

It was a busy place of ships visiting — cruise liners, cargo freighters, the *Wave Ruler* tanker, research survey ships and the Royal Navy. We had met government officers and British Antarctic scientists at King Edward Point. We had made friends of yachtsmen and museum staff at Grytviken. It was a busy island with a party life in full swing. There was no remoteness in South Georgia with hydroelectric power and endless buildings and machinery from the time of the whaling stations.

Roy was on notice, the yacht was packed tight, and we were away — never to return!

One thing this journey had taught us was how hard the goodbyes were.

South Georgia Memories by Morgause Lomas, then age 11.

Chapter Thirteen

South Georgia, bound North

Soft water exit. Power into the back of the mast.

"*It's the little ones, growling debris that form our greatest danger.*"

"*Our dreams were difficult in this faraway place…*"

"*There was repetition and reaction to focus the resilience of effort.*"

For our South Georgia exit, the weather file grib data was excellent. The first task was to escape the Southern Ocean icebergs of the Antarctic convergence zone. The cold South Pole water swings north of South Georgia by about 250 miles, and beyond this lies warmer South Atlantic current and no ice. Big icebergs are no great problem, easy to sight in daylight and scan well on the radar; it's the little ones and growling debris that form our greatest danger. There is no substitute for eyeballs. Every big one on the radar is given a good few miles berth. That's daylight to go then.

For our exit, we awoke to a soft snow covering for five a.m., decks white with ropes forming shadow lines that focused the mast.

Jean-Yves from French Sail yacht *Hinayana* cast our shore ropes away as we moved into the central bay of Grytviken to pass King Edward Point. The mighty whaling station cast its shadows to dance in the sunrise. The VHF was checked out with the British Antarctic Survey team, and there was one last salute for the grave of explorer extraordinaire, the boss, Shackleton. He brought all his boys home safe, and you can't argue with that.

With an hour of motor, we left Cumberland Bay and the lee of the mountains for open water. The ocean washed the decks clean of snow, and a third of the sails grew to half by ten o'clock. We were in the hands of nature, sailing on the wind.

Escorted by a huge sooty albatross to signpost the way, South Georgia's mountains set into the distance of our rear-view mirror on the Southern world of winter chill, and our minds were already focused on the warm arrival in St Helena.

It was snow, sleet, then sunshine all in the day. Never mind thermals, it was colder than that. Two pairs of woolly tracksuit pants under the heavy sailing bib and brace trousers had us stumbling about like giant, overstuffed teddy bears.

Good miles, good wind and good progress of about 70 miles as the southern winter sun looked for a place to set. There was only one big iceberg three miles to the port side. Caitland spotted it and bagged the name "CBerg" for the mighty white fellow. Why do you only spot bergs when the sunshine fades? Sunset was at 18.30, and our position local time was 53.00S 36.00W. North of 53 south, it must be getting warmer, but we could have done without that berg.

We were still moving, but slowly for in the darkness with much of the sail in and a stout lookout on the short-range radar scanner for an evening of icicle-chilly watch. A persistent flow of large bergs followed through the darkness to sharpen the nerves, focus the mind and turn the fingers blue. Dare you touch anything steel, it would peel back a layer of the glove from your hand as the cold froze it fast. With all the layers on, even to sit on the teak cockpit shelf for fifteen minutes would turn you stiff to the bone.

We had 52 south on the dial, and we were flying on good wind. How we felt the warmth of the sunshine and breakfast for that second day. Have we been too far south too long? We got a position email to Roy, put the sails back to half of our canvas, turned a little more west and ran well in a good wind.

Icebergs arrived three at a time, dancing on the twelve-mile radar scanner. Not a lot by local standards, eyeballs Caitland then watched them in.

Disaster Strikes

At 10:45 on April 14th, we were 180 miles out. The yacht rolled off course and circled around in the wind. The pit of your stomach aches in a moment as the yacht leaves its comfort zone.

It was instantly obvious that we had autopilot failure. We re-set and re-set again, but there was no rudder control. To a shorthanded yacht, the autopilot is probably the most important member of the sailing team; we call him "Ray", which is short for his manufacturers name. We tried everything on the re-sets, but there was no active steerage. We hove to. That's side to the wind with some reverse cutter. Our ahead speed was cut to zero, and we drifted down the wind at a few knots, but the direction of nature holds us steady as we are stationary with the water around us.

Hove to, we changed the core pack computer of the autopilot. Rudder feedback was OK, but still no steerage. A voltmeter was put across the computer output, and we found that the power to the clutch and power to the motor drive was OK. The physical rudder steerage is done by a large electric ram in the most rear locker. The rear locker is a damp space, great for storing the spare sails, deck chairs, beach toys and kites, but a big job to empty at sea.

Hours later, the sails were in the cockpit, and the beach toys of equatorial summer were resplendent in the winter sunshine of the Southern Ocean. Locker clear, there was access to the ram. The best chance was that the rose joint had pushed away and needed replacing. No luck today. The joints and drive bar all look well. We put a voltmeter on the feed cables at the ram end and there was power for the clutch and power for the motor. That told us there was no cable fault. That was bad; the ram itself must have been damage.

"R" clips out, rose joint cleared as fast as a racecar pit stop, and the big black ram unit was out. We took it back to the cockpit for a good look. We removed the

CHAPTER THIRTEEN: South Georgia, Bound North

tiny cover screws, holding each one like a newborn baby as waves gently rolled our ocean workshop. Inside the ram housing, the stationary side of the clutch had come away from the bulkhead. In this movement, it had ripped the cables from its electromagnet core and destroyed itself into tiny bits. Our hearts sank as we looked at the magnet carnage.

The drive ram wasn't a year old; we had replaced its ancestor after four years of good work. Back to the old boy. We steadfastly refitted and reconnected the best parts from each ram. The cables were connected and locked it in.

We had a sea trial there and then, but to little avail. The old ram had been tired, and that's why we had replaced it. We had swapped the drive belt and spleens from the new unit for this special occasion, but sadly the strength was no match for the Southern Ocean seas around us.

Tired in the long day, cold biting our physical efforts and disappointment eating our motivation, it was time to stow the locker gear as best we could. We fired the generator up and got some hot tea. We emailed Roy with news and position and got some sleep. Our dreams were difficult in this faraway place.

Breakfast time on April 15th, and we had a locker to repack. Success with the autopilot was not to be. We had tinned spaghetti Bolognaise for a yummy morning feast to cheer us up with some deep south motivation. Memories of that "Lady and the Tramp" film never do let you down.

The whole locker was sensibly re-stowed by ten, and that's going some when the kitchen table's not level, but it's fair to say there was no sand on those beach toys today.

Twenty days to Cape Town with hand-steering, and having to heave to every night made it 40 days. The best foot forward was in the form of a retreat to South Georgia. We'd had one good day out of great wind, but it would be three hard fought days back. The Southern Ocean winds are mostly all from the west. Sailboats motor, but hard, long-reach, wind-driven seas don't return Mediterranean speeds. Here it is like the Bay of Biscay on steroids, and mountain climbing by Snowdonia standards. Those black Welsh mountains were all around us today.

In balance, to hand-steer for Cape Town seemed beyond reach; the autopilot electromagnet was unfixable, and a short, hard run into the wind for the safety of South Georgia and a parts delivery from one of those visiting ships had clarity and sensibility.

The weather window was open, and if we raced all speed with some good evening luck on the hove-to sleeps, we would be back at Grytviken before the next big gale rolled in from Cape Horn.

Something about reliving those goodbyes helped the decision and returned the Velcro attachment. Down to it then, motor on, cutter sail out and some short main to make back into the wind.

Wind, sea and sail were cutting through, but it was hard steering. We took 30-minute sessions then a break for muscle relief of Popeye spinach. It was bitterly cold and the ocean was unsettled. It seemed like endless turns on the wheel to dance down and over the bow-breaking rollers, each and every wave moving us as far from course as nature could enjoy.

Six knots of success at 220 degrees, and that's going some against the sea in the screaming fifties. Hand-steering into wind determined to spit you back is an endless effort — like driving a car with a flat tyre.

Cold and tired, it was still a good day's work. By sunset, we were down to 128 miles from Grytviken for the day. Thirty-six miles was good for seven hours against the wind. We hove to with a reverse cutter to stall the wind and a full right rudder with the wheel tied against the harness bolts in the cockpit floor.

Thirty-five-knot winds were building into the darkness. People talk about slick water and softness when hove to, but when the seas are big, it's no walk in the park. Generator set on, hot water, you can't go wrong with a steamy cup of tea at a time like this. Bedtime eyelids were heavy, but the seas were growing ever more. We set three-hour alarms for iceberg checks on the radar. It's not like the ice has gone away in our moment of need.

We broke our own rule!

Never move a furler in the dark. The wind was well over 40 knots by midnight, and we decided to shorten the reverse cutter. Our ocean tactics were always to run and never stand a gale. To run down the wind here and lose ground was not an option.

This leg of our travels would bring great expertise in heave-to tactics, but we had no luck tonight. We had difficulties in the darkness as we shortened the sail, a flailing rope tangled in the cutter sail; the force of the hydraulic furler winding in the sail ripped the spigot on the furler foot. The centre was free to turn, and the sail unwound to sit full out. The slack ropes of the cutter sail then beat the decks of the yacht at the pace of our racing hearts.

We had little choice in the darkness but to secure the sail full. With the growing wind, we had taken a step back to finish our toils with more sail than less. How that drains the energy.

We pinned the ropes tight down on the cutter sail to make the best of a bad job. We used the winch sets, tightening the rope by hand to pinging tight — so stiff the ice-water spray bounced off the twists of rope. By three, the wind was clocking over 60. It was one of those jobs had we never touched, it we would have been better, but hindsight is a valuable commodity in the land of adventure.

Sunrise on Friday April 16th could not come soon enough. We had recorded a staggering top wind gust in the darkness of 88.2 knots true wind.

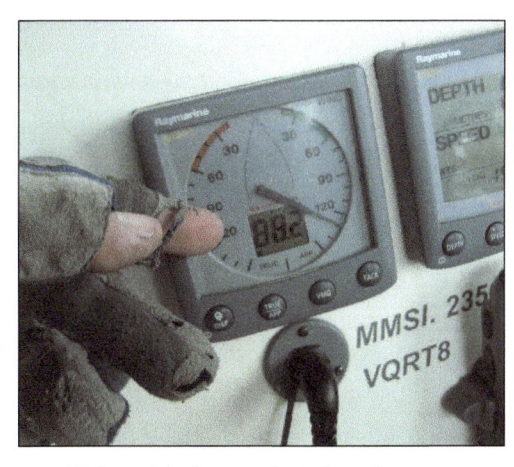

Eighty-eight knots of wind on the gauge

Hove to, we had drifted twenty-eight miles good, ground progress was made. Grytviken still lay 126 miles away. By straight line, we were two miles closer, but on a steeper wind angle at 250 degrees. Don't fret, we had a weather forecast grib showing a wind turn that was going to help.

The sea was tormented with the dark wind of the night, but we celebrated sunrise with a Pot Noodle breakfast, fighting nature to boil a kettle without spillage as we bounced from wave to wave. The cooker was on a gimbal, and there were edge traps on every kitchen top, but when the seas got this big, it was academic. You balanced on one leg, holding everything; you never need a teaspoon to stir it!

Heated on the inside, we cleaned the diesel filter for a healthy motor. Water in the fuel tanks from condensation in this icy land meant each and every job added to the effort when times got this tough. One last job, all that wind in the darkness had broken a dinghy stay on the portside davit. Buses come in threes and all that. Ropes meticulously doubled up, they had held the stay well, a quick cord and re-cleat to replace the strain of the broken stay, ten minutes in the sunshine, but an age with blue fingers sticking to the ice-clad steels. Gloves were torn up as their top layers remained stuck fast to the ice-cold metal.

Day two of hand-steering began, dedication and determination, our predicament was worthy of note to the big boys. Roy relayed our position and steering issues to Falmouth Coastguard, UK, and World Maritime Safety had our number on the distress board. People have often said, "Why Falmouth?" Falmouth is the world centre for any UK registered vessel.

With all the jobs, we had lost much precious daylight, so it was a late start. We were sailing around ten but very organised, flask of hot coffee by the steering, flapjack at the ready, and motor on again to maximise every moment of manpower and muscle effort. By eleven, we were closing a course of almost 260 degrees for six knots, making use of some close-to wind on the cutter. The cutter remained full, we could tack it, and we had it so tight the music of the wind was fit for any rock band. Go, go go for Grytviken. We were religiously set to 30-minute steering shifts, judging the forward rudder deviation angle to steering by the inch.

Wave watching and stabilising for best of best steering, the concentration was wearying. Steer for 30 minutes, drink quarter of a cup of warm tea, eat a sugar flapjack and be ready for the next 30 minutes. Toilet about every two shifts, there was repetition and reaction to focus the resilience of effort. Icebergs had become a secondary problem as we counted every inch as we watched that Cape Horn gale close on the computer updates.

If the gale caught us, it would shut the door of our return to South Georgia, sweep us not just back to where we came from but tighter into the berg zone — failure was not an option.

We hove to with sunset at six-thirty. Day done, enthusiasm running amuck and propping up the energy. Grytviken lay 76.9 miles at 245 degrees. Yes, point nine is important at times like this. What a day of elation, and we were much more positive on the position email to Roy. With each data posting came the replies of motivation and weather grib updates. We couldn't do this without that boy in Blighty. That wind shift for a turbocharger run was expected for the morning, but the evening wind was still 30-plus. Gale blowing by Blighty standards.

Friday evening was full of dreams of drift. There were no jobs left for the morning — the fuel filter was cleaned before bedtime to be sure. By nine p.m., the winds fell to a steady 25, and the roll was sensible enough to sleep with legs wedged across the bed for stability. The alarm was set for every three hours to scan the icebergs around us. Hove to, we all drifted much the same, so safety was good.

Our friends on the *Pharos* had often talked about rogue bergs. Sadly, one iceberg seemed to run the whole other direction, wind and current with ice was a strange business of logistics. There is always one!

By midnight, the drift was steady and holding very south as our wind turn forecast grew. The door for Grytviken was looking good.

Saturday morning was full of madness and excitement. We rose in the darkness for five a.m. to be ready for six sunrise sailing. It seemed futile to hand-steer into the wind on a bouncing sea we couldn't see — this was a daylight-only effort.

Steamy Pot Noodles for breakfast, flask full of coffee, and hot kettle water full to supply the first few hours push. Sunrise sensation. We had drifted 21.6 miles. Gryt-

CHAPTER THIRTEEN: South Georgia, Bound North

viken was 70.2 miles away at 260 degrees. We were more than six miles good. The drift distance was lower, and the angulations not terrible.

A day run was feasible; it was within our grasp. We had thought to target a southern bay on South Georgia to back up arrival, but this was magic. How we dreamed of warmth. This was fried bread with ketchup for breakfast or real butter on crumpets for teatime. We could smell the success in our tummies. The cold and the ache of tiredness was now suppressed by the excitement of winning a goal.

The very first Saturday daylight, we fired up the Perkins metal sail, turned the main and cutter out of the evening's stall and made every moment of the daylight. We counted those hours of the morning as a soft wind shift aided the sails. Breath held with anticipation, we steered up tight as we could to the wind and watched what we got. Balancing speed for angle, nature gave us 275 on the dial for six knots — go, go, go!

For all you salty sea dog navigators, best not forget we were running magnetic bearings here, and deep south leaves us about 15 degrees deviation from true north, so 270 on the dial is a good bit short of true west. For those of you off the sailing pace, the reality is the compass clock has changed by an hour with the magnetism of the Earth, but these readings were good for Grytviken.

Nine-thirty, the kettle water coffee was gone. You can't measure life better than that, but with 53 miles to go, the adrenalin was kicking in. With less wind on the bow, we brought out half the jib, more main and started to see six and half knots flash on the ground speed dial. Eleven o'clock broke the seal for the first of that flask coffee, 42 miles to go, pushing seven and a half knots for 275 on the dial. Apollo Thirteen rocket motors, Houston, we are good to go.

Closing two o'clock, the winds falling and land closing. Yes, we spotted the mountains! How many times can you divide miles and speed in your head to predict arrival targets? Every point one of a knot re-calculated the numbers. If you watch that kettle, it's always going to boil slower. Five p.m. and the target was coming true. We predicted daylight arrival for the quayside.

No, we were not there yet!

The mighty blue Polly Perkins diesel stuttered and died. No, please no! We were too close. The fuel filter was OK, so we had our hearts in our hands. As close as we were, if we missed this wind window, we would still be spat back.

A quick pump of the hand primer cracked the injectors and she burst back into life as unexpectedly as she had stopped. Maybe some rogue air from the morning filter change had caused a blockage; it certainly gave us indigestion. A darkness deep in our minds told us to beware, celebrations should not come before the win.

Three o'clock in the afternoon to see an eight-knot ground speed as the glorious metal sail Perkins roared hard and healthy again against the water, wind even settling to let us back, but the ocean still roared white and frothy and there was an eight on the Beaufort scale.

With the fall of that wind came fog. Our precious view of the mountains was gone. We edged closer to Cumberland Bay, every eye on the water ahead for growling iceberg debris. Hold the rail and look right out the front till your hands turn the gloves blue, then step down, thaw for a moment and return to duty. Well done Caitland and Morgause! The depths were shallower as our home in the water beckoned. The first signs of floating kelp were two miles off, and then we could see rock again.

The snow-clad mountain ring around Cumberland Bay looked positively warm.

Safe back into Grytviken

One day, one evening and a morning out, three days hand-steering, three evenings of drift, and 82 knots of wind on the gust dial. It had been a torment of the mind for numbers of distance and time, but the quayside lay before us. Rowing for regatta glory at Henley, this was a cream tea in the bag.

Government officer Robert was on hand on his Sunday off to take our ropes in Grytviken.

Three hundred and fifty seven miles on the water log. That's 180 out, 177 back plus 72 miles of drift. No, it does not add up, but that's nature, and the ocean isn't still. Like walking up the moving stair of the fairground funhouse, the return was no

CHAPTER THIRTEEN: South Georgia, Bound North

easy monster to tame, but God must have been on our side. Well done, Father Paul, and the blessing of the seas 15,800 miles back in St Katherine's, London.

The adrenalin and elation eased back. Our bodies were shattered, tired, cold in the bones and aching in the arms from that steerage. We had frozen eyeballs for the berg viewing. It was a hearty helping of Bolognaise and a bucket of chocolate for bedtime back with the fur seals of Grytviken. They almost seemed friendly welcoming us home.

Get a grip, all that effort, halfway in timescale to Tristan de Cunha, and yet we were still 54 south, and as much as the ocean was cold, we had forgotten the cold driven in the wind from those towering ice mountains of South Georgia. How is it you can never sleep when you're that tired?

Email to Roy: "Stand down we are in." He joined our sleepless nights with not a moment's watch missed. Well done him! You can't do these things on your own.

Sorting the problem out, there were no flights, and European parts distribution networks were all grounded because of volcano ash from Iceland. No ships for four weeks. Was our luck going bad or what? That's another story!

Chapter Fourteen
South Georgia Repairs

Back with the penguins, KEP Point, South Georgia.

"*Time and tide…*"

"*How eased the mind is, when a decision is made.*"

"*Our hearts torn with the enormous warmth of family on the ocean.*"

The European summer was warming the northern world, and general election dates were set in Blighty.

Hollinsclough lay in South Georgia, a crewmember missing, her autopilot was gone, and the southern season was turning winter. Life was not looking so good.

The irony of our return was our last words to David Scott Cowper of explorer motor yacht *Polar Bound*. "*Look forward to your next lecture.*" We had no idea it would be back in South Georgia!

A broken autopilot, the linear electric ram drive was all ripped up; the spare was old and unable to push the rudder at any speed. In the security of the quayside, we opened up the damaged unit to find the clutch magnet had pulled its location pins out from its bulkhead. Free to run, it span round, ripped its cables out and broke up.

The first step was emails for parts from Blighty — Iridium satellite communications around the world. Well done technology of talk when you needed it. Roy co-ordinated with Andrew from Cactus electronics back in London. A replacement existed in one of the Raymarine European distribution depots. It was good news, but there was no luck. A huge volcano eruption in Iceland had grounded flights across the world. A mighty dust cloud of jet-engine disaster lingered high in the northern skies.

Hollinsclough lay safely on the quayside again next to explorer motor yacht *Polar Bound*.

David sorted the girls hot chocolate and then helped take a look at the autopilot ram. The clutch was stripped out, and the boys from the British Antarctic Survey base at King Edward Point rebuilt the magnet from the old ram to the new one and got great resistance on the circuit. We then put the clutch of the old worn unit in the new ram.

Some success, but the centre screws would not tighten without freezing up the slider bearing. Raymarine Europe technical engineers emailed advice. Some more messing with the sticking centre bearing, and Loctite glue was left to set on the screws.

David on Polar Bound has hot chocolate waiting for the girls.

Boys and toys, it's a good repair, and this ram just may work.

It was a reasonable repair. David and I stood triumphant on the ice covered boards of the quayside with the ram and motor re built.

Things were looking good and a weather forecast grib file showed the big gale was to pass and exit in six or seven day's time.

Snow on the ground hampered an external tidy up — snow for April in Southern winter. Soft flakes, it was too dry for snowmen building but OK for snowball fights. Chilly Billy evenings, hatches locked down and heating on, run the engine for some more warmth and keep that window closed. The heating was off in the night, but golly, we had condensation turning to ice inside for the breakfast call.

Those 80-knot Southern Ocean winds had taken a toll on the rigging steels. A broken dinghy davit was re-bolted back to a fine holding bar, and the cutter sail furling spigot was withdrawn. It was cleaned, re-cut to shape and locked back in for a perfect repair. The girls had greatly grown up with the adult interaction of South Georgia and the hard-fought run for sanctuary after the broken autopilot.

Morgause was in training for the position of deckand and was doing a fine job with a hacksaw, pliers and rope handling.

Time for the lecture then. Everyone on the island of South Georgia gathered in the scientists' dining room for a night out.

David did a fabulous Powerpoint presentation of his polar trips. North Pole via the Northwest Passage. Seas iced across, lunar landscape of munchy, crunchy bergs frozen in. Icebreaker escapes, diesel deliveries by helicopter, and then a year on the ground. Caterpillar digs a hole and drags you out for repairs, that's explorer boating for you on an epic scale. We felt humble and proud to be neighbours.

Evenings out in Antarctica with a tasty curry to finish, yacht *Wanderer III* — Teece and Kiki — and we finished the day with a French yacht *Hinayana* recipe of best chocolate brownie cakes.

In those last few days of South Georgia quayside, the winter was clearly arriving.

CHAPTER FOURTEEN: South Georgia Repairs

The world above was circling clouds of wobbly curve lenticular spirals. This was a sky the like we had never seen before. God was reaching out to touch the ground. The white blankets of snow had transformed the austere shadows of the metal whaling station into a Christmas-card vista of delight. Himalayan-like mountaintops iced the cake, and with darkness, the Milky Way swathes of twinkle focused on the Magellan clouds in the dark sky of bedtime brighter than anywhere we had seen in the world.

Winter was arriving

With Christmas pudding and lashings of custard, we rebuilt the energy — it's all low calorie in this temperature. Fisheries vessel *Pharos* arrived, and Bob on the deck crew took another look at that dinghy steel. He took Morgause in hand, and together they spliced and swathed by hand to create a strength of steel stronger than the original. *Hollinsclough* was perfect again, and somehow Morgause was a gown-up sailor.

Time and tide, parts via the Falklands were possible, but it meant a month more winter to wait for the delivery ship visit to South Georgia for the last leg of logistics. To wait or not to wait, security of the repair was balanced against a new part and the cold Southern Ocean winter.

The skipper of the *Pharos*, Chris, got the big charts out and set all our minds to the best drift, current and wind in the deep oceans. Decisions needed to be made, and Chris' expertise and experience were invaluable to our decision.

Girls on board or not? Byron Marine offered an evacuation berth on the *Pharos*. Falklands school friends in Stanley, and Tamsin's parents, Mike and Margaret McLeod, offered to baby-sit. We could sort a military flight for Brize Norton after a short term of school back in Stanley. Maybe re-route to Cape Town from Heathrow. It was a difficult decision to split up or run on the ocean as a family.

We formed a new plan of action. Parts were needed for a proper repair of the autopilot. The European Raymarine distributers were still struggling to source the ram unit. The original target was making Tristan de Cunha and then north for warmth.

It was ten days sailing, 1,500 miles to the windswept Tristan islands. Still good, but from there, we decided to re-route for Cape Town. That's another 1,500 miles and ten more days, but engineers were a plenty, and suppliers and distributors by the bucket full.

South Africa, southern winter, but warm by Blighty standards, and home to the Royal Cape Town Yacht Club. A 50-tonne crane too, that's got to be the way then. Did we mention the football World Cup?

How eased the mind is when a decision is made. It looked good, and the weather window was getting ready.

Meanwhile, we had a few days left on South Georgia.

There have not been many 13th birthdays at British Antarctic Survey King Edward Point station.

Caitland's 13th birthday loomed, April 28th, she would be at sea on *Hollinsclough* bound for Cape Town. A quick secret discussion with base commander and former Girl Guide Ali and a special Friday, St George's Day birthday games party was secretly planned for five o'clock. What fun can a girl have so far away?

Musical chairs, champion George, treasure map, teenage maths test, weigh the cat, number of toffees in the jar, test your taste buds. Jenga balance blocks — move as many as you can in a minute, one hand please.

Morgause grew as a deckhand, Caitland celebrated her teenager date, and the scientists had a ball.

Ali won the play-off Jenga championship with 20 blocks. The wettest game in the room was bobbing corks, lifejacket and goggles on to grab them with your teeth. A fun set of games to score points, Richie bagged ten special extra points for a spiky hairdo. Paula ten more for making the birthday cake. Wow! What a birthday cake. It was shaped like a king penguin, lemon markings for the golden chin. It was a South Georgia celebration feast fit for any teenager.

Just like Blighty, a girl has to have a few dates for her special day of 13 years. Hav-

Party with BAS scientists

CHAPTER FOURTEEN: South Georgia Repairs

ing had a magic Friday games party on the BAS base at King Edward Point, Sunday was set for a repeat performance on the *Pharos* with Chris and his team. The girls had become *Pharos* family with the big red fisheries ship.

Morgause and Caitland set the games in perfect order. They used the comfy mess

Pharos ship party for a 13-year-old

room sofas to start the fun of musical chairs. The aft poop deck was used for a team trial on the wet-wash cork teeth grab. What a giggle as Carl added snow and ice to the water. Then it was back indoors for the table games. There were some variations, like a Spanish language puzzle to find a Christmas tree. The treasure hunt map is always a success, and Gogo toys replaced toffees to find 120 in the bag. Name the Ice Age movie character was no problem for Stu before checking out the maths test. A perimeter of triangles and circles, key stage-three maths for teenagers, it was a cool game. Chief Engineer Kim checked the numbers but still no extra points for the engineers as the deck crew pulled ahead on their points. It was neck and neck first to third place over 80 points — what heroes? Paula from BAS was chief photographer for more than 300 memories and Doc Susan adjudicated for a little more practice for next time's win.

There was nothing for it but a play-off on the musical chairs. It was a good job the chairs were soft for the heavy squeezing and sliding-in moments with the vigour and enthusiasm to get bums on seats first. The scores on the doors found Boson Carlos the winner. Matty G and Robin were one point apart for the runners up. Skipper Chris provided the prizes, a very proud pair of *Pharos* crew T-shirts.

Chef George delivered the dinner. Mighty great Chilean steak sandwiches with thin-cut chips and lashings of dressings to be devoured with a fresh-fruit pudding desert.

Family together, we togged up for the walk home around the bay from KEP to Grytviken. Howling winds had built after teatime. Birthday party over, we waved goodbye to our *Pharos* family.

They beamed out the iceberg spotlight, and we raced our glowing shadows in the snow. We walked home like stage stars of the Royal Albert Hall in rings of sunshine to warm our hearts for bedtime dreams of fun and games on the *Pharos*. The

Big Red Ship was a warm memory that will last Caitland a lifetime of birthdays to come.

Monday morning and a new week, charter yacht *Seal* left Grytviken and Cumberland Bay for the south anchorages. *Wanderer III* took the trip with them. That same day in soft air, Ron returned from the very same south anchorages on his American single-hander *Relentless*. Soft air days are changeover days in the island life of this faraway place. Turns in the wind beckoned, and it was time to pack, re-pack, and stuff even tighter for the run ahead. Motor yacht *Polar Bound* set all east for Cape Town.

When does the wind come? We checked the weather forecasts gribs half daily. There would be no fish and chips on Friday for us. It was time for another go at escaping South Georgia, but the weather said no! A change in grib forecast placed ever more unexpected wind from the north inbound. Not a good day then — maybe Saturday. No, again steadfast on the ropes. A north wind brings sunshine and snow in Grytviken; we washed and cleaned the yacht topsides while BAS scientists helped a seal trapped in a fishing net escape its tangle.

May 1st – Pinch Punch, first of the month

It was time to remind our Spanish-speaking chums of the English phrase "Pinch punch first of the month." We sent an email to Girl Guide Guia Scout friends in Santiago, Conception, Valdivia, Puerto Montt and mainland Chile. What a tale of travel to tell folk which themselves had been so much part of this trip.

The message was: "*Hollinsclough* waiting in Cumberland Bay, South Georgia, Southern Ocean waters of the South Pole for a good wind to the remote Island of Tristan de Cunha and then fair wind to Cape Town."

May Bank Holiday Monday May 3rd it is. The gribs looking excellent, near south wind for midday to push us out and away from the mainland, lots of west to follow with only a single day of north in the seven that looked good for heaving to. Roy Dunning organised a berth at Royal Cape Yacht Club Cape Town.

Snow was falling in Grytviken, Island of South Georgia, as *Hollinsclough* dropped her ropes and slipped away from single-hander Ron on his American yacht *Relentless*. Museum manager Ainsley waved goodbye.

We eased across to King Edward Point for a snowball fight with the big red fisheries vessel *Pharos*, skipper Chris got a photo from above, snowballs frozen and floating in the water!

CHAPTER FOURTEEN: South Georgia Repairs

A flat calm exit for South Georgia

We waved to the scientists of BAS and motored into Cumberland Bay on flat calm Southern Ocean water.

This Velcro really was torn, winter was in South Georgia, and there was no coming back this time. Our hearts were torn with the enormous warmth of family in the ocean. With the friendship greater the farther you travel, the goodbyes just get harder each time.

Chapter Fifteen

South Georgia – a Second Exit, bound Cape Town

Snow building in Grytviken, South Georgia.

"*I pulled hard and broke our favourite red winch handle.*"

"*By a miracle of Newtonian gravity, I kept myself upright…*"

"*There was no horizon. The dark cloud covered sky joined the edge of waves…*"

May Bank Holiday Monday, May 3rd, leaving South Georgia a second time, Caitland had become a teenager in this remote icy, mountainous island in the Southern Ocean.

Our hearts heavy with goodbyes, *Hollinsclough* was well rigged. Well trained in the prevailing wind, we set her spinnaker pole to starboard and ran a jib rope through the end block for an easy, quiet, downwind run. It was not the most efficient way of sailing, but we could run silent in the evenings, and sleep on these runs had proven a valuable commodity.

We had seen 80-plus on the wind gauge returning into South Georgia, so we set the yacht battle ready. A gale is no place to be stowing and setting rigging when you are shorthanded. Just for good practice, we moved the second life raft from storage to the cockpit floor. It made a sort of doorstop under the table and brought a warmth of safety. The boat was very well packed, and all doors we wouldn't need were taped closed. And various spare bedding was stuffed into door areas to wedge every last thing that could move.

The autopilot was on from the start; we were very keen to see it hold. Good wind and away, cutter full, half a jib, and a third of the mainsail. Steady morning winds saw South Georgia disappear at about 15 miles of visibility.

There had been no rush, matching a great weather window to get the very best of the forecast in the next few days, we just used that first afternoon to get ourselves clear. During our adventure, we had rarely used the practice of stalling the yacht out for steady sleeping in the evening, but our autopilot failure had taught us the practice of heaving to. Energy spent on the cold was high on our list of value we needed to bank. Awareness of the bergs and the value of sunlight travel for watch were critical.

For that first evening, we hove to 45 miles out at sunset. Jib in, cutter two thirds in reverse, straight tight and six feet of main unfurled, full right rudder strapped. We had it textbook and timed it to the very last of the working light.

Fishing vessel *Argos Georgia* was working the area, her lights bright. Caitland radioed the skipper in Spanish. "I have you on my AIS," he said, all was well.

A rolly evening, we drifted ten miles good, and then as the wind shifted, we lost five to awaken 52 miles out. It never adds up, but remember the sea is moving, and for a night's sleep, we were in the very best shape. We stuck to our trusted watch pattern, six hours on and six off, split midnight to six a.m. Tracey was on the first darkness, I took the midnight to sunrise shift. We had the radar scanner on twelve-mile scan and the added safety of a deep-water trawler watching over us. We would have loved some more favourable drift, but it was a good night and the next day's weather looked great.

Tuesday

Tuesday sailing started from hove to around nine in soft winds and sunshine, but it was cold. People talk about cold, but deep down in the south, you do adapt astonishingly quickly. The cold this morning was OK, but best not forget, the ropes were frozen on the winches. Touch a piece of steel, and it would stick fast to tear a layer of glove off your hand. We were always worried damaging a frozen rope; we eased them free of the winch sets one loop at a time. There was not a sign of icebergs. We had good visibility, and the ocean still enough for accurate wide radar scans.

We had a great day on the sails. The jib was stretched tight against the spinnaker pole, and we found ourselves 105 miles out. Steady progress, low miles to show for the work, but we were ultra-cautious about icebergs. We had had great visibility, and we were dealing with the cold well. Our energy levels were very well balanced, and as the day ended, we were quite chuffed with our progress.

De-rigging the sails, the darkness just overtook our work on that second evening.

Taking in the jib, the block came free of the spinnaker pole. The block is just a turning wheel making it easy for the rope to run free where we fasten it to the pole.

With some excellent help from Caitland in the darkness, we re-ran the jib rope conventionally to the main winch and set the ropes conventionally. We recovered the rope block, not sure if the spring catch had broken on pole or the lug itself was damaged.

It was a lost opportunity of tight jib for downwind runs, and it drained some of the motivation energy from the bank. There would be lots of opportunity to have a look at pole in a better sea state and daylight, but these things spoil an evening. Was it an omen of things to come?

We checked the rest of the pole ropes — all was well and secure. Our heave to was even better for the evening, as we had a touch less cutter and got it very straight. We found it best to be drum tight, so we hand-winched while steering bang into the wind. You could have bounced a ping-pong ball off it.

We played with the main to get the yacht very beam side to the wind, making her steady hove to, about six feet unfurled, full rudder and we even strapped it both sides. Each watch sleep was good tonight and surprisingly warmer. Three-hour checks on the radar at three, six, and twelve-mile scans saved electricity with shutdowns between. There were no icebergs to be seen, and that offset the sadness of losing that spinnaker pole block. We drifted at good speed to claim 31 miles in the darkness.

Wednesday

We were 131 miles out of Grytviken for Wednesday morning and day three. It was another sunshine day of good wind with some south in it. We set the jib conventionally, about a third of it out, moved the tracker to the middle and left it fairly tight. We set a full cutter and a third of the mainsail.

Progress made in the daylight was about 70 miles. We were just short of 200 miles from Grytviken by darkness. Southern winter days only give eight till five — it sounds like an office job.

Darkness arriving, we got an iceberg two miles starboard of our drift target. We popped the motor on and pushed for an extra 30 minutes. We positioned the yacht north of the iceberg in the last of the daylight and had three miles of clearance. All was well, but it left us to draw the sail in the darkness.

With the loss of the light, we had a second evening of bad luck. This time we tangled the jib. It is really straightforward work, but we must have misjudged the effect of that cold. I have never been sure, but maybe we ran the furler the wrong way. We unwrapped the jib, drew loose rope and re-wrapped it, but it tangled. Whatever, it was a mess, but it was safe, and it was time to get ourselves half hove to.

We set the cutter to two thirds of sheet out. Here came the next omen. To get that drum tight effect, we always hand-winched to feel the strain on such a tight rope. I pulled hard and broke our favourite red winch handle. Damn, it was tight, but I never knew you could break a winch handle. Some hard worn spigots had broken and been discarded, but in all our miles, we had never broken one before.

It touched our hearts so deeply that evening that we had a ceremonial chucking of the broken handle. With the handle gone forever, we set six feet of main, watching the drift come very square. The rudder again was fully strapped both ways.

The darkness was no indigo, the motivation down, luck lost on the furler, and absent of our most favourite winch handle.

The job was not done. The yacht sat beautifully, but it was still a big sea bouncing around from top to bottom, and that front jib sail looked very wrong in its middle.

With a harness on a tight strap, it was time to venture down the jack line in the dark and take a good look at the jib. It just was not one of these jobs to be left until morning; if it unwound more, we would be up in the night for a real fright. "Let's get it done."

There was disorientation in the darkness of that big sea, but the sail was fairly tight on the base of the furler. Looking farther up into the darkness, there was trouble that twisted the stomach. A balloon of cloth high up sat like the knurl of an old twisted tree.

The whole evening was noisy with a big wind screaming. Close up on the furler, it was clear the jib itself was very noisy as that bulge fought the wind.

I remembered climbing a rock ledge in Snowdonia and thinking I was glad the fog was covering the scary view down below. Tonight, the sea state was hidden in the same manner by the darkness. All things relative, I just focused on the deck pushing against me and twisting my balance. The waves out of mind, I heaved and pulled the sheet ropes, but to no avail, and by a miracle of Newtonian gravity, I kept myself upright.

The deafening wind and crashing seas left no option but to make a return to the cockpit to communicate a plan of action. Back down the jackstay strap and into the cockpit. With mind to unwind the sail and re-furl it, Tracey took station on the hydraulics. Morgause was on the deck rail to shout against the sound of the waves, and I returned to the bow.

Returning to the base of the furler, I tried a winch handle on the manual to little avail. Shouting in and out, relayed by Morgause, Tracey operated the hydraulic motors for the furler. After a number of tries that seemed to take half the night, some sail cleared. I was very worried to see the sail number oddly low. I had great fear that we had torn the jib sail.

Without calling back to Morgause, suddenly there was lots of out movement on the furler and a particularly rogue gust of wind blew the jib sail wide open. The whole sail beat down the wind. First, it was a great relief to see no tear of any kind. Then it was time to take cover from the ferocity of the whole front jib sail. "In, in, in!" I screamed back to Morgause. Bless her cotton socks, she had a clear view of the intensity of the sail in its entirety and screamed back to Tracey with the same intent, "In in in in in!"

The force of the wind pulled the sail tight and straight from the furler, the hydraulic motor rolling it in, wrapped it beautifully. It was a huge relief to see the sail was well and watch the mighty force of the hydraulic motor beat the wind.

CHAPTER FIFTEEN: South Georgia – A Second Exit, Bound Cape Town

I was now lying on the deck ten feet back from the furler, securely harnessed to the jackstay line. As the sail rolled in, the sheet rope securing it was now long and loose and beat down on the deck all around me like it wanted to cut the yacht in half. Morgause tried to grab at the rope as it thrashed from the winch set. One return beating of the rope struck me across the shoulder and spat the torch from my hand. It went down, light blazing into the darkness. It seemed to twinkle beyond its distance, but it was gone forever. The rope on the short side went tight, and the job was done.

Tracey came off the furler button unaware of the rope battle taking place on the deck. Morgause's screams were deafened out by the raging water. In a moment, she pulled winch side of the beating rope, and in that action, the world around us was secure.

I bodily rolled over the lip of the cockpit side to fall onto the teak seating base. Morgause took grip of my harness for comfort, and we all smiled at each other with the same words, "We did not die today."

Golly, we were tired. Caitland was on hand for a prayer to God: "The sad loss of our favourite winch handle and best torch." The attachment to such kit in these environments is astonishing in the value of its motivation.

The wind gauge was plus 60, moving around bow and stern of the beam as the yacht bounced about. Stalled out and hove to, it was still thrown about beyond all imagination of an anchorage. There was one more trip to the stern to fire the generator on, and it was time for the microwave to provide the feast of a hot Cornish pasty tea.

On any trip to the stern, it was always a joy to stand full starboard rear holding the aerogenerator stay and take in the might of the seas. In the reflection of the jib adventure, tonight my eyes then took in the size of the ocean. There was no horizon. The dark, cloud-covered sky joined the edge of waves, climbing around us like a tunnel.

As those pasties steamed in the microwave, we did the Wednesday night posting to Roy.

"Going well 18 local Wednesday. Sorry to be a bit late. Hove to for the evening, but we tangled jib reefing in and had a hell of a time, but all well again. Sadly lost my favourite torch. 51.56.50S. Yes, 51 land 32.33.00W 197 miles from Grytviken. 1226 to Tristan. 65.5 miles made good today, most all on the wind, no diesel used so that's good too. Three days sleep deprivation settling, all feeling worn out, but it must get better from here. Think tomorrow is soft but following day turns, any advice on making north or west good on motor for tomorrow. Carl."

Thursday

Thursday morning and sunrise greeted us without cloud. We had logged a fabulously, favourable darkness drift of 24 miles.

Last night's iceberg was still a healthy five miles to our rear. The 50th parallel south beckoned 100 miles north.

Clearing the ice convergence zone to the South Atlantic would find warmer ocean and safety with no icebergs. Tristan de Cunha read 1210 miles on a direct line; it seemed around the corner, and the ferocity of the tangled jib sail a distant memory.

The Thursday forecast was no wind. Motor on, the course not all north but about 30 degrees. Pot Noodle breakfast, and things must be getting back to normal. There was a scattering of icebergs, and Tracey named one "the football boot".

At twelve o'clock, 16,000 miles came on the log. Gosh, 16,000 miles of adventure from Blighty. The sea had gone flat enough to have the heating on without fear of damaging the diesel blower.

Woolly hats and all togs on from breakfast to bedtime, we never removed them even for sleeping, but it was warmer that day.

With 50 miles of motor for the day, the batteries were full and the fridge cold. The eye of the storm was ending, and the pressure was falling.

We hove to at five-thirty, 275 miles out of Grytviken for a soft stop and hot soup with white bread Ainslie had baked back in South Georgia. Yum scrum! We had made the 50th parallel south territory with 50.53.30S 31.22.80W at 18.00 local.

The contrast of the day was unimaginable.

The forecast ahead was poor. We had been aware from the off that Friday was our bad day. What was forecast as a 12-hour monster was turning into a 36-hour beast, but it didn't weigh on our minds. The yacht was well, solidly stowed, and we had crossed our sleep deprivation to be at one with the environment around us.

Hove to and aware of the turning winds, we had shortened the stall out of the sail, and it was tight.

Half a reverse cutter and just four feet of main rolled out from the mast, and full left rudder tied down both sides to be sure.

Tracey took the six to midnight watch. I slept well with not a single layer of clothing removed. From bed to cockpit, all that was to be done was clip my harness and life jacket straps on.

With midnight, I looked up the stairway to see Tracey above in the cockpit. The seas were building. The notes in the log were more scrawled with the movement.

CHAPTER FIFTEEN: South Georgia – A Second Exit, Bound Cape Town

The wind gauge pushed 66 knots and would move forward and stern as the yacht twisted in its stall. We exchanged watch with a hug, rolling about like overstuffed teddy bears.

It was a cloudy night, the view black as black can be. There were no stars and no moon to be seen. All that motoring left the luxury of electricity, so the radar scanner was permanently on that night. At one a.m., there was nothing on the scanner; we had the luxury of three, six and 12-mile checks.

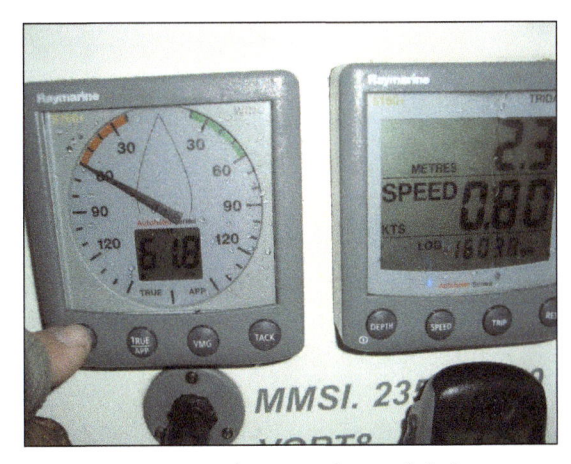

Sixty-one knots on the wind dial

Waves around us making more than half the mast, we gauged them at plus 30 feet.

With the wave state huge, even hove to I was cautious we were rolling so hard I set the scanner to maximum offshore on its sea state to be sure it knew the horizon was poor. Mountains of water half the height of the mast only let us have a view of the distance from the top of an occasional wave set. The majority of time, we were rolling, slipping and sliding down the sides of the peaks and dales of nature's wildest landscape.

I was wedged tight in the cockpit, short harness strap to the U-bolt by the engine instruments. While the yacht was being tossed about, I was well locked into its motion. I still took occasional views over the cockpit screen, holding the steel top rail. Whatever was out there was ragging, and it was a big one.

At two a.m. almost to the clock hand came a sound so unusual it scared me to the bone.

The sound was the first sensation, and I found myself stood bolt upright holding the top rail, knuckles so tight water squeezed from my glove edges. In my mind, we had hit something head on. To my astonishment, as I looked across the forward decks, there was nothing. It was an impact, but nothing was there.

The sound of crashing water came down over the roof of the canopy like a reef breaking close in on the shore. It was the frightening sound of imminent disaster.

I swung around from the rail, but there was nothing on the radar scanner. Instinctively, I flicked the switch for the mast spreader lights to illuminate the deck. We never used them sailing as they blinded out the farther distance in the darkness.

I got a clear view over the forward deck, its teak glistening in the icy spray. As much as it sounded like water was showering down, there was nothing there.

Morgause screamed out from below decks. In big seas, Morgause found more comfort sleeping on the saloon sofa. She was starboard side and had been thrown across the floor in the impact. The shock awoke her screaming. Tracey arrived in the cockpit stairway from the aft cabin, but refocused on Morgause.

For me, time slowed down. I turned around and, holding the small winch head, I leaned out over the cockpit ledge to get a good look to starboard. The water looked like a milkshake.

There was no doubt we had hit ice. Starboard there was a showering spray of ice debris everywhere, like a snowstorm that bit you with its edges sharp.

The water had turned white. It wasn't a cresting wave, and it churned in a most peculiar manner, like an egg whipped in the mixer. My ears were full of the thunder of sucking and slurping that rang out louder than all the ocean around me.

The white monster of froth and madness was very low, entirely wrong in its motion; I may never have spotted it even in the daylight. The spinnaker pole had no block, but it was still set starboard. The rope triangle ties often touched the water in these big seas, but they never caught on this monster.

The favourite winch handle lost, the best torch gone, and this evening the circumstances that followed could not have been more unlucky. We were textbook sailing for safety; we were hove to and stalled out for security in a huge sea. Tonight this action of being stalled out was our downfall. Everything relative to us was stationary; the monster was going nowhere soon. It slammed, slammed and slammed again as it munched down the yacht. The milkshake water took on a rage of intensity in the huge sea as it slowly swept down the whole starboard side.

The time scale of this affair was the age of a lifetime and lasted forever in what must have been about a three-minute window. We rolled against this hidden danger more than 20 degrees. I was now stood on the starboard deck holding a stay line as far as my harness strap allowed me. I peered farther into the darkness. The sound was stronger than ever. It slammed and fought the yacht behind the cockpit as the froth of tormented ocean glistened in the deck lights. Then there was one last heave. We rolled hard port, and it was extreme. In an almighty gloop, the monster sucked down under our stern. The crashing sound ended in an instant, and the only sound left was the rage of the big ocean.

Hollinsclough lurched back upright to return to the normality, and the clock in my mind returned to normal time.

CHAPTER FIFTEEN: South Georgia – A Second Exit, Bound Cape Town

As I took a breath, I turned to the cockpit. Tracey had calmed Morgause, and she was beside me. All that was before us was a mountainous dark ocean without mention of what had just happened.

Oyster build their boats well — *Hollinsclough* was hardly creaking and taking this world in her stride. We were shaken, but she was not stirred. None of the bilge pump warning lights were on; we appeared to be intact. I strapped my harness to the jack line and took a look down the deck. There was not a rail or guard line out of place. The seas were huge, by the bow I was on all fours to check the anchor-mounting rail, but every pole and rope was fine — nothing was touched.

Below, Morgause and Tracey pulled some floor panels up and checked the centre bilges. To some surprise, all was well — all pumps off. Below decks, rolling in the motion of a roller coaster, they continued. They pulled open cupboards low in the galley and the rear walkway floors to stern — not a single sign of water.

Caitland remained fast asleep in her bow cabin. The only change to her world was that she fell asleep in one bed and would awake in another. She must have bounced across the cabin in the final heave, and by another piece of the growing good luck for the evening, she was flung from starboard to port without a bruise.

With such bad luck of the last few evenings, we had had the luckiest escape of our lives. *Hollinsclough* was no racing machine, she was heavy and strong, and the strength of her English build had saved our lives.

Morgause calmed, the ocean was so big there was no safety in moving around, so she snuggled up with Tracey in the stern bed, and I returned to my watch. My mind was twisted with the event. Tracey got some sleep; we would need her energy for the next watch.

Six a.m. Friday, and I handed over watch. These were big seas, and it was difficult to move around the yacht safely. I took a brief look under the centre bilge floors — all was well. It was a warm sound to hear the clasp of my harness open. I hung it on the bedroom wall and wedged myself sideways to Morgause, who remained in the depth of a slumber. The harness swung on its hook like a pendulum as it fought to find gravity while the bedroom windows displayed washing machine spin beyond their toughened glass.

Tracey logged a red sunrise at seven-thirty. "*Red sky at night, shepherds delight, red sky in the morning, shepherds warning.*" The winds were building 50 plus. She logged a 66 on the highest gust as the wind came about from the north. The sea state was Southern Ocean roller coaster white and wild.

The plan of action was to ride the storm stationary, stay hove to for the day, remain stalled out too lose as little ground in the wind turn, and return to reality tomorrow when the forecast changed back. When it was bad down here, the gauges would go off the scale, but with such anger, it was always over soon. An odd thing that would contribute to our future, we never once unstrapped the steering to move the rudder.

I took watch early around eleven. Tracey had entered the log at 10.30 local reading 51.20.00S 30.36.75W hove to moving three knots 145 degrees. Tristan de Cunha was 1161 miles away.

We had zero over ground on the forward gauge, but the GPS gave three knots on the drift, losing all the north we had made yesterday, and we were a whole lot more east. The winds were over 50 but falling, and the pressure levelled.

For the start of my watch, we appeared to be through the worst of it. The forecast of 30-foot seas was spot on — walls of white were all around us. As we climbed up each big set of waves, it was clear daylight, and there wasn't an iceberg on the scanner at any range. Eyeballs Caitland took a rare look topside at the world raging around us and confirmed that her view was clear.

Late after lunchtime, both Tracey and I were in cockpit. Tracey was forward facing behind the steering wheel with an excellent view of the computers. I was in the starboard forward seat, harness clipped in above the engine gauges.

The big sea state was calming, nothing over 20 feet, and the yacht had been sitting steady in the gale. We had chatted about the forecast, 12 hours of gale was looking like 36 as the low pressure anchored over our heads. When was this weather going to let us go?

I had the most unusual sensation, and the air was spat out of me.

My harness straps were so tight they tried to squeeze the life out of me. I glimpsed the most peculiar sight of the second life raft passing above me.

In a blur of sensation, I refocused. I felt like I had been hit against a solid wall. I was looking outward at the ocean. I was hunched over the winch head, deck side of the cockpit lip.

I turned around to the cockpit and calmly said, "Tracey, you need to get out of the water."

There was no feeling of cold, and the sound of the ocean had emptied from my ears.

She wasn't in the ocean! The cockpit was flooded; it was full to the brim of the rail like a giant bathtub.

The later log entry would read:

"Slammed down 14.45

"Friday May 7th Approx 14.45 local rolled flat on our side while hove."

CHAPTER FIFTEEN: South Georgia – A Second Exit, Bound Cape Town

This was probably the moment I broke my collarbone.

Full togs with multi layers, it was luck that I was so well strapped in shape by the clothing. There was much disorientation in my mind, but as I found myself sat back on the upper rim of the cockpit facing back inside, I remember so clearly telling Tracey to get out of the water like she was a naughty girl going for a swim.

There was nothing more spoken between us while the water drained through its flood grids in the cockpit floor. Short breaths were highlighted by the steamy mist of the cold. Much of the dark blue cockpit canopy was torn down around us. The table remained upright, and the chocolate wrappers of an earlier snack were gone.

Tracey and I sat there without words; we were both thinking that it was the luck of God that we had all the washboards in. The sea state had been calming; had it not been for the cold, the companion way washboards would have been out, and all that water would have poured inside. We had been slammed down so deep into the water we may never have righted.

The water did not drain quickly; it was rare luck that we had the lower cockpit window that gave air to the galley kitchen closed.

With much of canopy torn down, the first priority was to take cords and lash. The blue sheet was flailing around and ripping more of the roof section out.

There was nothing like a job to aid you through an emergency. Instinctively working together, we quickly got the rear sides rolled and strapped and then lashed some cords between the steels with a few extra guy lines.

Pain arrived, but it was hellishly cold, and we were so wet it all seemed normal.

Battered, bruised, disorientated and even a little unsure what had just happened, we sat for a moment, almost leaning back to relax in the war-torn cockpit. Smiling between the steamy breaths, the sensation of life was warming back into our souls.

Full consciousness returned to our ears. Down below, Morgause was shouting, "There's water coming in! Lots of water!"

Tracey was first downstairs to calm Morgause.

As I turned for the stairway, my eyes were caught by the engine gauge panel. Warning lights were red, and the bilge pump was on! No! It was not just one pump, but all of them!

My mind torn, that was a lot of water to drain, but how on Earth did we fill every bilge?

Descending the stairs, all the lower kitchen floors were flooded, and the most peculiar thing of all was that there were blobs of oil gelled up in the water.

It took less than a moment to grab the small silver T-shaped key and twist the two locking bars, removing the engine room panel.

Water then rushed out everywhere, only dammed by the bulkheads. Polly Perkins was deep, and she was flooded above her starter motor.

Caitland was still in the stern bedroom, and she shouted, "The rear heads are wet!" No one was having a shower, that was for sure.

Beyond the bedroom door, there was a frightening dampness about the air. I could see the floors were awash, the rear quarter floorboards to the head didn't need lifting, they were floating! Caitland, wedged on the edge of the bed by the aft shower, was ankle-deep in cold, icy saltwater.

I was braced in the doorway; the yacht was still being kicked around, the seas at 20 feet. Caitland was well wedged in, but she looked at me perplexed.

The water was draining; it all ran forward to the big pumps in the centre bilges and carried the engine bay oil as it went.

The boat was floating OK, and we had taken cockpit floods before. We had left windows open in the past and soaked the place, but never before had we seen water like this.

There was silence in the recognition of disaster. Safe as we all stood there, there was a whole lot more water than there should have been. My mind ran to the cockpit flood, the drain pipes run right through the engine bay and exit the lower hull. I pulled on the large hose, but it was solid and tight to its fittings.

My mind woke up to the intense reality that this was not water from the cockpit flood.

My mind raced through the events. The knock down seemed a distant memory, but how on Earth did the yacht go unstable in a softening sea?

The ice debris growler impact came back into mind like a roller coaster car crashing from its rails, but the run of recent coincidences were longer.

The autopilot failure on the last run out of South Georgia — had there been impact damage that we had never considered? The sail back found the hand-steering work a huge effort, but then we so rarely ran off the autopilot that it did not seem out of order in big seas. The return across the flat water of Grytviken was heavy on the steering, but there was a perfect explanation when alongside the quay; we had found a rudder full of kelp.

So keen to push the autopilot to test it for the Cape Town run, we had hardly hand-steered a mile on our second exit.

Hollinsclough had a skeg rudder. Well protected, the solid skeg lies ahead of the hinged rudder, giving it enormous protection, but when we hit that growler, we were hove to and took the impact sideways.

CHAPTER FIFTEEN: South Georgia – A Second Exit, Bound Cape Town

There can be no doubt the ice growler ran the length of the yacht, bow to stern. We never checked the prop cutlass. Was that leaking the water in the engine room? Was the rudder stock twisting?

Much as our minds hunted answers to maybes, the real question was the lack of stability and the weight of water hidden in her stern.

Hollinsclough had weathered far greater seas and never been close to a knock down, but today she was flattened into the water. Was water building in the rear quarter bilges under the aft heads port or behind the fitted starboard wardrobes?

Was the lower rudder stock twisting into meltdown?

Chapter Sixteen
Realisation We are Sinking

Caitland's GCSE project, portraying the moment of impact.

"*The ocean was entering our home without an invitation.*"

"*Reality was the next clarity…*"

"*I think we need help…*"

Late into the afternoon of a softening gale in the Southern Ocean, we were astonishingly at one with an environment more hostile than anywhere else on Earth. Our minds crashed through questions. Battered and bruised, we looked at our log entries for the last half hour.

Log entry:

"Friday May 7th approx 14.45 local rolled flat on our side while hove."

"15.00 South Georgia local time 51.32.40S 30.11.90W Drifting hove to at 4knots approx 90 degrees."

Rolled flat was no light entry to put in the log. It had never happened to us before in any sea of any size, and we were not even sure it could happen while stalled in the wind and hove to?

By that second three o'clock log entry, the cockpit was clear of its knockdown flood, but below decks was a small war zone. The whole yacht was ringing damp with steamy cold mist everywhere, but the centre bilge was holding fine with the pumps on and draining down the water from the stern and engine areas well.

It was now clear we had water ingress from the stern — the ocean was entering our home without an invitation! Water must have been flooding the stern rudder lockers, and when they filled, it was clearly finding its way down the yacht by way of a route under the bed and rear quarters. Water steadily drained forward as each bilge filled and then finally over the engine bay to flood into the centre bilge to the main pump.

An answer to one of the questions from our knockdown was clear: the flood that followed was from hidden water built up in the stern and not from the bathtub flood in the cockpit.

It was clear we were carrying a whole load of water aft that was building up to cause the instability that brought the knockdown. That water must be the result of the growler impact in those early hours of darkness. It was so close in timescale, but that was a trauma we had gladly shelved into the back of our minds and replaced with a luck of God escape.

There was no system to drain the aft bilges under the rear quarters of the back bedroom. The rear cabin was fully enclosed — wardrobes, drawers built in from new and all very full of kit. The aft external locker forward of the davit posts was deep, and it was stuffed full, two spinnaker sails and a pair of rolled-up dinghy shells amongst many other items. Halfway into the depth of the locker, a sub-floor protected the rudder bars. This was no sea to empty this locker, and certainly no sea to try and get that sub-floor out to look farther.

Had the knockdown damaged the rudder area further and increase the water ingress? It was a question we dare not ask ourselves, and then there was a clatter on the windows, making our minds flinch.

It was the weather saying hello as a deluge of rain came crashing down. The ocean was still wild but smoothed, and it took on a surface layer of goose bumps with the weight of the rain. As freezing as the rain was, we warmed to the signpost of a front that would see the low pressure gale pass over. The winds recorded as low as 30 knots while the sea state was becoming very confused — heaving, twisting and turning, it was hiding the instability of our hull.

Bobbing about stalled out to the wind in the turning ocean, it was soon after the rain we took two more poopings of the cockpit as rogue waves swept over the stern and landed topside. They cleared as fast as a roller hitting the beach.

Hollinsclough held upright through those rogues. Our hearts pounded with them, but we were so relieved not to have been knocked down again. The floods drained down the engine room pipes, and there was no repeat of Morgause screaming out, and no oil over the floors.

We hadn't tried the engine, as there was a lot of water down there. "Let's save it and not risk melting the starter with a huge short on the solenoid."

We were a little cautious of the electricity.

While there was lots of water in boat, the navigation electrics were all good. The pumps, having cleared the first flood, were no longer permanently on and had begun cycling, only kicking in to clear water as it raised the float switches in the bottom of the centre bilge.

Pot Noodle or cup of tea? The kitchen cooker systems were all awash with damp.

Tracey and I took turns to change clothes in the aft bedroom cabin. We were both sodden to the skin and cold to the bone from our swim in the cockpit. While getting into dry clothes, I could hear a lot of water in the stern locker.

As I walked forward into the galley, Caitland and Morgause were at war with the kitchen water. They had begun a tidy up, and they were sweeping the floors with towels and ringing them clear into the sinks. A cup of tea was on the horizon.

Standing in the narrow galley, it became apparent that we had running water somewhere behind the fridge systems and the aft wardrobe lockers of the bedroom. I started to realise we had a much bigger water ingress problem.

Standing behind Caitland and Morgause, I pushed my arm past a fully packed cupboard of provisions to the side of the fridge. A slatted run of teak is the last layer before the fibreglass hull. My finger ends twisted between the teak slats and the fibreglass felt very wet and oddly soft.

CHAPTER SIXTEEN: Realisation We Are Sinking

I moved back to the computer desk. "16.00 position 51.34.07S 30.06.14W, 285 miles northeast of South Georgia. Friday evening weather settling, wind down to consistently low 30s, sea state still confused and fierce."

By the time I had the log entry done, Caitland had two steamy mugs of tea waiting.

Back in the cockpit with Tracey and two steamy mugs of tea, our motivation was up with a hot drink in our hands, but we were out of earshot of the girls, and it was time to take in some reality.

We were both back in dry clothes but not warm. The cockpit had changed. All the starboard canvas had been torn down, and the wind and ice spray now ravished us with bone-deep chill in every gust. The tea had to be drunk with speed if we were to savour its warmth.

The icy gusts through the canopy gaps were evil, but the weather was settling for sure. We were getting some big washings, and the ice spray lashed the seating area, giving even more discomfort. It was not a discomfort as strong as the discussion we needed to have about the adventures of the last few hours.

Good fortune broke that discomfort, and we recognised that we were reasonably stable. Also, there were no bergs on the radar.

Reality was the next clarity, and I spoke in a hush, "*We are definitely taking water in the stern rudder locker.*" Between us, we were clear that there was no opportunity in this sea state to explore that stern locker. It would be hopeless to excavate gear in these seas, and it would be dark very shortly.

The pumps were holding the bilge levels, but it was plain scary to see that much water. We looked at the pump-on warning lights in the engine gauge set. They were cycling, but they were on most of the time before our tea was finished.

We had a discussion on tactics and hatched a plan of action for the sailing. We had to take all opportunity to motor and sail to get north in the morning's better weather. We couldn't see that we could make any distance back, so we had to go on. We had to escape the ice water of the Southern Ocean convergence zone, and it couldn't be more than 50 miles north of us.

Maybe we both knew we were sinking, but it was good to change the subject to sailing and add warmth to the truth.

With the arrival of sunset, we lashed the upper canopy a lot better. With energy levels with the knockdown getting low, I pressed Tracey to try and get some sleep. I penciled in a double-hour radar watch for bergs as the darkness fell around us.

With so much coldness in the cockpit, I returned below every ten minutes to warm and took the time to log the email on the Iridium satellite.

I sent various emails to Roy: "Boat not well but berg problem greater can only push north. We can hear a lot of water in the stern locker but unable to access. Second life raft in cockpit area. Await any advice." I sent similar emails to our family of friends on the *Pharos* ship and also to South Georgia.

We got an email in from Mike routing weather from the Falklands, and he advised an extension of the north wind gale. It was softening, but it may well be with us 12 hours longer than expected.

Back in the cockpit, there were no icebergs on the radar, and much as we are being kicked around, we clearly had a good mast horizon for the scanner.

In another ten minutes, I was downstairs again for more warmth. I pulled some centre floors up to find that the water was way above the pumps; it was no longer possible to see the float switches.

I reluctantly returned to the narrow cupboard by the fridge. Reaching deeper, I got three fingers tight between a back slat. It was a rare piece of fibreglass hull I could touch. I sighed and a darkness tunnelled my vision. My fingers felt like they were touching soft butter. I could feel the tactile strands of fibre, and they were squashy, not solid. I pulled out much of the provisions in some anger. In the angle of the back of the cupboard, there was as much as a square foot of the slating. There was no damage, no gushing water, it was just soaking wet. It was seeping everywhere I could get torchlight on, and it was all soft. It had a mouldiness about it, like a going rotten feeling.

We had worked this yacht halfway around the world, and I knew every bit I could get at.

Forward of the lounge space, the core packs of the hydraulic motor system sat close to the hull. I pulled back the sofa tops and cleared some spares wedged in the side of the core control unit. I could see six inches of hull top and eighteen inches of metal core box. It was the same sight, no damage, all in shape, but everything was soft. There was seepage in every section I could see. Pushing my hand against it, it was soft; I could claw against it with my fingernails.

I sat back at the computer and typed an email to South Georgia, but I didn't send it: *"I think we need help."*

I woke Tracey. She was huddled with the girls in the aft bedroom cabin but not really asleep. I took great caution not to raise undue alarm.

We both looked into the bilge, and the water swirling around with unhappy motion was ever higher. Tracey was shocked at how much it had come up since she looked. "Are the pumps on?" she asked.

CHAPTER SIXTEEN: Realisation We Are Sinking

I sat Tracey in front of the computer, and without words, we both paused in our own thoughts. I pointed at the email and said, *"We have to send this."* She gripped the mouse with a dry hand and clicked "Send."

The next email was to Roy. "Please call Falmouth and advise we may have a problem."

We both took a look around the yacht. Water was in every bilge. I pulled some board in Morgause's starboard bedroom wall. I could get at more slats, and I found the same fibreglass, soft and soaking wet. With the board out, I could reach all about the hull. There was nothing broken, no hole or damage, no sharp edges, but it was wet through and appeared to be seeping. I checked the forward heads, portside it was easy to reach past the toilet exit pipe to the hull. The fibreglass portside was cold and damp but as solid as a rock. Portside was as good as ever and a huge contrast to the starboard fibreglass.

I sat back with Tracey by the computer, speaking softly, *"That ice debris growler must have torn up the outer laminate of the hull from bow to stern. The damage outside under the water line must be far worse than we can imagine."*

A sound we had never heard before startled us.

The sound was a blurping softness, and it took a moment for us to focus on where it was coming from. It was the voice handset of the Iridium satellite phone system.

Our whole communications system was based on satellite phone link, but for data email, not voice. Data is very cost-effective and efficient, as it gets low error rate anywhere in the world. Voice on that system has a big time lag when switching and can be very broken up. It had a handset, but we had never used it as a phone.

We pulled the handset free of its mount and with trepidation held it across both our ears.

The call was from Robert in response to that email five minutes earlier. Robert was the government officer on South Georgia, himself a former sea captain with a wealth of deep-ocean experience. I ran our situation past him, tears in my eyes, explaining the damage.

There were great gaps in the sound as the signal switched voices and a long silence before we got back "Is the email a mayday?"

I'm not sure what my answer was. I think this may have been the greatest challenge of my sailing communications. It's very hard to actually say "mayday."

Sat together with Tracey on that sat phone, there was water everywhere, and damage from bow to stern as the hull seemed to be delaminating. We admitted defeat.

"*Yes, it's a mayday.*"

Robert was very clear. "*Swap from emails and use your EPIRB.*"

That's the little yellow transmitter that every ship in the world carries. If it gets wet, it automatically triggers the position, presuming the vessel about to sink. You can pull them out of their holders and trigger them manually. With a long timescale of sailing, we had accumulated three, and we had all of them aboard *Hollinsclough*.

Email: "18.00 Position Mayday 51.35.20S 29.58.07W Drifting hove to at 3 knots 80 magnetic."

We activated our three EPIRBs.

Three EPIRBs fired, one had GPS and whacked me in the eye when its aerial popped out. Morgause chased the plastic cover across the cabin floor. No lights. A good old-fashioned shake and we got one light. We taped it to the steering rail outside and got a green light. The oldest one had no activity. We tapped the last big yellow one on the rear back stay as high as we could reach, and then it burst into life with a flashing beacon.

The mayday logged, the EPIRBs triggered, we sent Roy an email on tactics to be rescued. We even thought about the tactics we were taught in our sea survival course back in Blighty at the chilled pool of the SBS Marines.

Email to Roy: "We have the spinnaker pole boom rigged on the starboard side of the boat. Two life rafts on board ready for despatch. If the yacht is floating, we will stay on it. Will email every hour while power systems hold. Flares on board. Epirbs tapped upstairs for best signal. AIS functioning. Radar scan still working and berg watch OK. Will only try engine if we have to as all awash with water."

There was a message in from Mike on the Falklands regarding a fishing vessel that was three days away, and he told us the *Pharos* ship was almost back at the Falklands.

The screaming 50s wind run out of Cape Horn for Cape Town would have been busy in the days of sail and whaling. It is a great wind route but commercial vessels today have motors and run way to the north on the shorter course.

Our own AIS location for vessels registered nothing, and Mike's email was no real surprise that the nearest vessel was three days away!

Deep 50s, Southern Ocean, South Atlantic. We were alone.

We faced the reality that we were going to sink, and no one would be close enough to rescue us.

The ocean temperature was well below zero, and survival in a life raft would be measured in hours in that cold. With the rate the water was coming up the bilge sides, we calculated about 20 hours of flotation.

CHAPTER SIXTEEN: Realisation We Are Sinking

Back in the Falklands, you had to sign for a permit to land in South Georgia, and I remembered the notes so clearly that stated, *"No means of rescue is provided from the island."*

Hollinsclough would be sunk by midday tomorrow!

We would be dead soon afterwards.

Tracey and I had focused on emails and the phone call, and with various floor panels open around us and water high in the bilges, we had failed to notice Caitland and Morgause alert and awake and in the galley listening to every word.

It was Morgause's voice that refocused us from the darkness of looking into the computer screen at our impending doom.

"If we are going to die, can I try a glass of champagne?"

"Most certainly not, you are not old enough," replied Tracey. "Why not? We're going to die" added Caitland quite aggressively, then family together, we burst into giggles.

With the carnage of a broken yacht around us and no possibility of rescue, we focused on family.

We togged the girls in full kit and got some cold pasties out to eat. We had not given up yet, food and fluids was imbedded into our ethos of sailing.

"Bother the cold pasties," said Caitland. "Let's at least eat the best chocolate we have been saving." She was not wrong; It was Hershey bar heaven for teatime.

I set a rota for one-hour berg watch as long as the electrics held for the radar, but it didn't seem a priority anymore.

In a life-affirming interlude, the four of us sat together on the cabin's large bed. Survival of the yacht lost to our minds, we hugged together and reminisced about our best adventures aboard our Oyster *Hollinsclough* — she was one of the family.

While we knew we would lose her, we celebrated her life, running downwind under the giant spinnaker of soft equatorial air, the beautiful ice glaciers of Patagonia, swimming with turtles and reef adventures with the giant fish, Abrolhos Bank with the humpback whales and Robinson Crusoe Island with unforgettable lobster tea parties.

Chapter Seventeen

An Asset Available, But It's Too late

Floor on the move as the the water rises in the galley.

"*My state of mind was out of focus.*"

"*Two children two toilets two pumps.. a match made in heaven.*"

"*A toil rewarding to the core of our souls. We were fighting for our lives and it was working.*"

Iceberg debris, a growler tears up the hull from bow to stern in a raging Southern Ocean storm during a two a.m. indigo darkness. The yacht survives, but 12 hours later slams down sideways into the ocean and floods. Back upright and stricken with damage, there is less than 20 hours of flotation. The nearest vessel for a rescue is more than three days away. in the subzero Ocean, we face no likelihood of survival. We face certain death, so we focus on family.

The blurping ring of the satellite phone broke us from our family warmth. How dare reality draw us back to our peril?

The dampness of steamy breath was all around us — no heating, no engine warmth, and water sloshing about the floor. It was with reluctance I stood up and went forward to the computer desk to answer the satellite phone set.

The sat phone call earlier, Robert from South Georgia, had focused us on a mayday transmission to recognise our fate and added no motivation to answer this one.

The voice was very scrambled, and I was unsure if it was from Falmouth Coastguard or the Royal Navy, and the only words I really remember about that call was, *"We have an asset available."*

Hellishly cold, it was fair to say I was a little confused rather than excited.

I was still sat at the computer desk as the phone rang again. The words I remember this time: *"England's finest available."*

With those couple of calls, my state of mind was still out of focus, I had absolute clarity there was no shipping vaguely in range. What on earth was this about?

By the next call, Tracey was by my side and the girls stood holding the galley rail behind. It was Robert in South Georgia.

"Warship *HMS Clyde* can be with you in 24 hours."

The ecstasy of the news was overwhelming.

"Absolute clarity earlier, there was no shipping vaguely in range." This had been confirmed by everyone in our communications chain. One group of vessels outside that chain was warships.

By incredibly unexpected fortune, *HMS Clyde* was 24 hours south of us. The warship was on a mission she hadn't even been tasked for, the luck of God was with us.

Emails came in from Falmouth Coastguard and Stanley FIC control.

Everyone was in different hours — Blighty was UTC, Falklands and South Georgia both askew to their local time, and our mind was in countdown of distance to target. It was hard to track the hours.

Our nine p.m. outbound log email sent to Roy read:

"21.00 South Georgia time, sorry lost track of UTC. 29.43.63W heading hove to 90 mag. Wind turning a little, presently west of north 20knts. No icebergs on 12-mile scan, stars in sky, clouds clear."

That was a fabulous course for Cape Town. Yacht log 16039: "So tired but a lot of water gurgling around the boat."

The log clear, we had lost connection between the time zones in the ecstasy of a rescue, but it didn't take long for all of us to add up that 24 hours was probably too late to save us.

The ecstasy was outweighed by pain, darkness was more indigo than ever, and the rescue was going to be too late!

Life in the yacht was becoming very tense, but now there was a cause.

The rest of that night, it was time to fight for our lives and get rid of some of this water.

Every pump in the yacht was on. The first crisis to focus on was that the water in the centre bilge was just about to wash over the top lip of the battery box. Twelve six-volt batteries providing nearly 1,000 amps at 24 volts were about to be flooded. They were the heart of our power system, and there would be no pumps without them.

That power was driving a pair of 2,000-litre-an-hour electric bilge pumps; they were running four tons of water an hour. While at first the electric pumps coped well with the damage, it was clearly getting worse as the rate of water rising was accelerating. Scar damage from the growler debris impact to the outside of the hull must have been eating its way through the fibreglass in a de-lamination process that was getting worse by the hour.

There was one centre bilge hand pump in line of the electric pumps. We had often used the hand pump to clear a blockage the electric pumps couldn't push through. It helped move more water. Tracey took the first stint while I thought about our other systems.

I remembered the trick told to us in Uruguay by Ian. He had been a friend of Baden Powell, the best boy scout trick in the sailing book. "For a big flood, turn the stop valve on the saltwater cooling feed to the engine, slice the intake pipe open and use the motor cooling intake as a giant pump." I looked back in the engine room to find that the starter was below the water level. There was no chance of an electrical circuit to fire up Polly Perkins.

Buckets didn't seem an option yet; with the yacht still being kicked around, nothing was level. The bilge to the companion stairway, cockpit and over the side was a long way. Everything soaking wet, we were struggling to stand up in the sea state;

CHAPTER SEVENTEEN: An Asset Available, But It's Too Late

floor panels were floating in the water, it was an accident waiting to happen that we could not afford.

Tracey was sensibly steady on the bilge pump. I stepped forward, looking at that fibreglass in Morgause's cabin. I turned around and faced the forward heads. The thought of the engine with its intake valve closed to work as a pump matched the toilets.

Flooding begins

Boats and toilets, that's the talk sailors always get to. Posh vacuum and electric toilet flush systems in the past had always blocked. Setting out in *Hollinsclough* 16,000 miles earlier, we wanted a failsafe system, so we had retro-fitted new straightforward hand pumps to the toilets. Flush was achieved by a few strokes of the handle, and the toilet was plumbed to exit overboard when at sea. In the same motion, it sucked in water from the ocean to flush.

We had two children and two toilets — a match made in heaven.

Morgause was already by my shoulder, and I stepped inside the toilet heads opposite her bedroom cabin. I locked off the intake flush valve on the hull fitting, sliced the pipe away that lay below the water and stationed Morgause on forward toilet pumping duty. It was clearly her station, next to her bedroom — next to the room that had been her home in the waves.

Caitland could see her task before I spoke. The other toilet head was in the stern, portside of the main bedroom. Caitland was in there before me. "Let me get the pipes sorted first."

The intake valve handle was near frozen, and I breathed a sigh of relief when it turned on the hull fitting. With so much water back here, I hardly needed to cut the pipe. With each roll, the water inside the yacht was close to going clean over the top of the pan. Success was solid in my mind, but I sliced way lower to be sure. Caitland was on duty, her task set.

I was back in the centre of the yacht to mark the water level on the battery box. Even in that first ten minutes, we had made a difference and held the level.

We had four tons of diesel on the boat, but I couldn't think of a way to get it out. There was half a ton of fresh water in the drinking tank, so I ran the sink taps dry. There was more than enough bottled water about for drinks. The sinks flushed into the grey water tank, and I used the grey water pump to clear that tank into the ocean. Then I thought, *where is that pump?*

The grey water tank pump was under the floor near the front head where Morgause was working with such an effort she was sweating. "Steady, take your time. It's going to be a long night." She was most disturbed to be moved. I lifted the floor by her side. The pump was perfectly placed to help; it was called a gulper, able to move some debris and ran at about 200 litres an hour. I cut the intake pipe short of the tank end, twisted it into the front bilge and left the pump running.

Morgause grinned as the pump rattled. She was happy to be back at her own work on the toilet pump.

I was aware we may lose electrical systems with so much water everywhere, so in a quick email to Roy, I advised we would focus all future emails to him. We would leave Roy responsible for forwarding to Falmouth, Falklands and South Georgia.

That was the last email. Iceberg watch was over, I took turns with Tracey, and together we put a consistent joint effort into the main bilge hand pump.

It was well past midnight when we took breath. It was joyous work with a cause — a toil rewarding to the core of our souls. We were fighting for our lives, and it was working.

It was a very vague calculation between shifts, with about half a litre a stroke on the toilet pumps and conservatively eight strokes to the minute, the girls had shifted 250 litres an hour out of each toilet. In three hours, they had cleared nearly one and a half tons of water through the two toilets.

Condensation dripping from the ice-clad windows, we had rarely been this warm in recent memory.

Saturday May 8th

Tracey and I sat on the centre cabin floor; it was going to be a long haul. We had turned the tide on the rising water, and I could almost see the pumps in the bottom of the bilge. I suddenly thought we had saved the batteries; we had better get some charge in them.

The big amp alternator on Polly the Perkins was out of the question, the starter motor and its cables soaking wet. I went outside to fire the petrol air-cooled gen set

CHAPTER SEVENTEEN: An Asset Available, But It's Too Late

in the top locker. Ice frozen on its rails, I pulled the hand cord with one hand, using my other hand to balance on the frozen deck. The first pull was stiff and gave nothing. With the second pull, it felt freer, and on the third it clattered into life. It was slow to speed up, but then it raced. I backed off the choke and let it warm. It was no time to stall and choke it before I flicked the power switch over for the electricity feed.

The girls were in the galley before I was back downstairs. To them, the generator sound was like an alarm. Electricity available, action stations in the galley. Kettle on to boil water, microwave fired up and hot pasties. "Steady on the power, one load at a time please, let the gen set warm up!" I shouted.

Back upstairs, the radar scanner powered up and an iceberg scan was run.

Downstairs, Tracey had the computer up and emails sent.

"01.00am South Georgia time 51.27.26S 29.31.43W Drifting 2knots at 70 mag Wind NW 15 20knots Stars gone, presume cloud, pressure stable. Yacht floundering around in what seems like a much better sea state. Fearsome bangs in the aft rudder locker are at every wave that hits a little harder. Water holding but floors waterlogged in aft heads, sound of loads of water in aft locker behind back bedroom. Waiting for daylight."

Tracey joined me topsides in the cockpit — the weather was a whole lot better.

Caitland was soon on the stairs with the first mug of steamy hot coffee. Two hot chocolates followed, and family together, we were all sat in the cockpit of *Hollinsclough* like we were on the leg of any other adventure.

The emergency bleeping of the EPIRB was a soothing sound, and with all the work of the pumping, the icy wind was a comfort to our toil.

I was cautious that we needed rest; we would need a steady effort of energy for at least another 12 hours. With such success on the water level, it was time to get some sleep. Mugs of hot drink empty, not a zip of clothing undone, it was back to the aft bedroom cabin for the four of us to squash into one bed and dream this time of a rescue.

Much as we tried, there was no rest from the sound of that water behind the bedroom wall. The banging of the rudder was louder and harder with every wave and outweighed the clatter of the gen set running above.

Caitland and Morgause were a picture as they fell into slumber, but Tracey and I were back for the next stint.

The generator was clattering away with delight, and I filled up the petrol tank as Tracey sent an email.

"04.00am South Georgia time 51.25.98S 29.24.18W. Drifting 1.6knts 80 mag Wind W 15, some stars partial cloud, pressure stable. Yacht still rolling hard, judders on rudder. All mid bilges holding."

Back to work on the pump, the night was gone in a flash, and how we welcomed the light of the sunrise through the cabin windows. The ocean around had taken on a calmness by Southern Ocean standards. The wind was down as low as ten knots, a blue sky flattened the waves for a wide ocean horizon that welcomed one of the finest new days of our lives.

Next email out was to Roy.

"7.30am Please get this through to Falmouth 07.30am South Georgia Time 51.26.78S 29.15.66W. Drifting 100 degrees at 1.5knts Wind N 12 Blues sky visibility good, sea rolling softer. All well, boat very damp and wet but bilges holding, spirits good all well, going to try some pasties for breakfast."

That's a huge rate of drift stalled out with the wind down in the ten-knot scale. The ocean was still well and truly wiped up, but all things relative, it seemed very calm to us.

Each time we logged the computer on, in and outbound emails would all activate in one data burst, and there would be mail to read.

The first email in was from the Vicar of Stanley. It gave us more warmth than the whole night's pumping.

> Dear Carl, Tracey, Caitland and Morgause. Sorry to read this morning on the BBC's website of your encounter with a growler:
>
> *A Derbyshire family are waiting to be rescued after their yacht hit a low-lying iceberg in the South Atlantic. Falmouth Coastguard picked up their emergency signal from northeast of South Georgia and alerted the Falkland Island authorities late on Friday. Watch manager John Rossiter said the vessel was stable but had taken on water and had no engine power. Warship HMS Clyde is making its way to the family and is due to arrive later as bad weather conditions improve. "'Growler' iceberg," a coastguard spokesman said. "What they've hit is a 'growler', where hardly anything is out of the water and the majority is submerged. You can track them by radar or visual lookout, but you can't see them all." A couple and their two teenage daughters are on board. They are not believed to be injured. Mr Rossiter said: "The Falkland Island authorities have now spoken to the couple on their satellite phone. The Falkland Island authorities are co-ordinating the incident and we are liaising with them and with our colleagues at Rescue Co-ordination Centre Kinloss." He added: "There were 50-knot winds from the west earlier on, but the weather is*

CHAPTER SEVENTEEN: An Asset Available, But It's Too Late

now improving. The warship HMS Clyde is 200 miles south of the yacht and is making its way towards the family."

God bless you with a calm mind and all necessary courage, and keep you safe. See you again soon, we trust. Richard and Jen, Vicar at Stanley.

That was a breakfast email worth waking the girls for.
"Where shall we go to the toilet?" asked Caitland.
How we giggled.
The biggest shock was that the world seemed to know more about us than we did.

From our darkest hour of certain death, we had turned the tide of our disaster, and we faced the new day with hope.

Through the first hours of that sunshine, each of us kept calling out, "The Royal Navy are coming to rescue us!"

I don't think we dared believe it, but the more we screamed it, the more it must be true.

Next burst of emails.

"9.30 am South Georgia Time 51.27.19S 29.10.92W Drifting 90 1.4knts. Wind soft 10knts N Excellent vis, boat rolling all over the place and it is low in the water."

There were a host of emails in, and family together, we jostled for space to join the friends who were all there with us.

Royal Naval Vicar Father Ralph on HMS York sent prayers.

There was a message in from chums on the supply ship *Shackleton* and the RFA tanker *Wave Ruler*.

Chris skipper of the *Pharos* told us to drop weight and loose the anchor chain, run rope and fenders off the bow to draw the yacht forward into the wind more.

Morgause and I always took station on the anchor windlass together, and with the email from Chris today, there was no exception. Easing the manual chain brake back, we put a good number of stops on the windlass so as not to snag the chain in the gearing. It squealed loud, but all was well as it left the yacht into 10,000 feet of ocean.

The pumps remained good for four tons an hour, and with a sensible push on the hand pumping, we had achieved time travel and were ahead of the flooding.

There was time double-check all the safety gear. We strapped the flare bucket on the top rail ready for action and then took a look at the life rafts.

There was no point in them sitting on deck.

We dispatched both life rafts on the starboard lee side of the yacht.

Tracey fired the first life raft. She had practised it so many times back on the RYA courses in Blighty. Morgause grinned, as it was her turn to pull the cord: "Emergency only. Dare I really pull it?"

There was something very warm about the safety of the rafts, their orange and yellow day-glo reflection all the more wild in the morning sunshine.

We set the large eight-man towards the bow and the special survival six-man to the centre of the yacht. We dispatch the danbuoy pole.

As we counted down the last six hours for the arrival of *HMS Clyde*, we ditched a fender in each of the following hours to leave a trail of our drift. We tasked the girls to pack what they could, essential and necessary, and the stuffer bags were filled to bursting.

When Caitland took family photos from their frames on the wall, there was a stark realisation that we were going to leave our home. Morgause dug out the skateboards from the workshop. "Daddy, we have used them all over the world, and we are not leaving them."

We were back at the computer for email coming in. Mike McLeod re-posted hourly weather turns more locally from the Falklands and never missed a moment.

Back in Blighty, Roy stayed awake for the whole watch so as to remain in constant contact with the Falmouth Coastguard, who were coordinating the warship being sent to rescue us. As well as Roy's vigilance, we particularly appreciated his messages of support and encouragement. One communication was particularly heart warming:

"Hold fast, a grey funnel ship is on the way."

ETA for *HMS Clyde* was two in the afternoon local time. She had steamed all speed ten on the dial against the weather. Skipper Lieutenant Commander Steve Moorhouse and crew were taking a pasting. The gale passing over us was slamming southern water swells direct into her bow.

By ten, she was posting a midday arrival.

Water across the floors was growing again. The mid-quarters, galley starboard and workshop port were all holding two to three inches of water. It sloshed about as we rolled. Water on the kitchen worktops had frozen. The navigation computers were excellent, inbuilt waterproofing, but the lighting systems were failing, diode bulbs flickering. It was no time for a disco, but first radio reception was about ten-twenty.

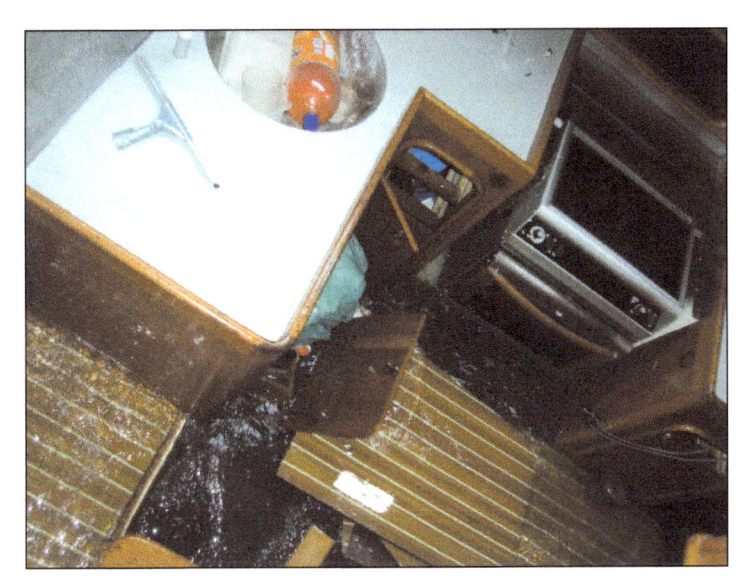

CHAPTER SEVENTEEN: An Asset Available, But It's Too Late

Water on the worktops had frozen.

We were so excited that the ice and cold had gone from our bones. We were not going to die that day.

All huddled by the cockpit handset, we could hear the warship on the VHF.

Our radio range was far shorter than theirs; we could listen but not get a transmission back. *HMS Clyde* was quoting our target position. Morgause got the coordinates down, and they must have been the finest numbers she ever entered into the radio log. They were spot on our position.

It was time to use the computer one last time and send data on the satellite. We confirmed, confirmed and confirmed again direct to Roy for Falmouth.

It was time to shut down the systems.

All floor areas were awash with water, and the battery box had started to be flooded. There was a nasty smell of battery acid. It's a sort of stink that's really uncomfortable on the nerves, a smell all wrong and against your instinct. This was no time for a fire.

The deck generator was still running, but when I checked the charge ratios, they had fallen a staggering 30 percent, shorts must have been all about the cabling under the floors. The charger may have been swamped as much as half an hour earlier. The high-amp primary master buzz bars sat in a seat head still above water. I turned the big plastic keys, both positive and negative, and the whole boat shut down.

The bilge pumps were off.

In the next 30 minutes, the water began rising alarmingly.

Bags packed, orange flare smoke filled the sky.

By eleven, with water growing, we had evacuated downstairs.

Family together, we huddled in a line on the aft deck above our back bedroom. We sat togged up for action in full survival gear, line straps out, dagger knives to hand for rope tangle. Morgause had her Navy hardhat on.

I donned a Royal marines T-shirt that had been a gift from our sea survival tutors. MOB bleepers adorned our harness straps, wet glove sets for ropes and woolly hats all round. Our best spotting bins were focused west. Morgause had a direction-sighted compass around her neck and still held the autopilot handset. That's habit for you.

Electricity was gone, but we still had a handheld radio on deck.

HMS *Clyde* asked for flares. Tracey got a VHF signal back while Morgause and I dispatched an orange smoke flare. Two parachute reds up in the next ten minutes, and then another orange smoke canister to fill the air around us with colour.

These flares were all in date, but more than half of them failed, and we tossed the duds in the ocean. If they had been fireworks, we would have taken them back.

The moment was Caitland's, bless celebrating her birthday back on South Georgia, she was a teenager and this was a moment for the best party of her life.

Yes, eyeballs Caitland bagged the first sight. No scream, it had always been her job to call the dial, and she just said it: "Smoke on the horizon. 275 degrees."

Calm as she had been, there was nothing else for it but the family song.

CHAPTER SEVENTEEN: An Asset Available, But It's Too Late

Sunday school teacher Shirley Duggan from St Marks had taught us well, sing it loud:

"All things bright and beautiful"
"All creatures great and small"
"All things wise and wonderful"
"The Lord God made them all."

The silhouette of the warship grew at such a speed that we could hardly finish the first verse before the detail of her mast and superstructure were clear. *HMS Clyde* is an ultra modern Star Wars machine. She had giant angular radar scanners and a very tall mast structure. Her bow wave was white and wide as she ploughed towards us.

HMS Clyde on the horizon

Roy's email, "Hold fast, a grey funnel ship is on the way."
Blimey it was here. We just sat there looking at it in disbelief.

Chapter Eighteen
Hollinsclough Sinks

Goodbye to our friend, she was a member of the family.

"Binoculars pinned to our eyes, we watched HMS Clyde come about."

"Lieutenant Rob Stavely, Royal Navy, how may we be of assistance?"

"HMS Clyde saluted with a long single blast of the warship's horn."

The Southern Ocean gale was gone, and only when *HMS Clyde* ran close on approach did we take in the reality of the sea state around us. *HMS Clyde* tearing down the waves of the Southern Ocean water showed that this was still no millpond, the sea was angry, disturbed and almost hostile. The wind was down, but the ocean still carried its force.

Binoculars pinned to our eyes, we watched *HMS Clyde* come about to take lee of the wind and launch her fast response RIB. How small it looked in the waves, and how the sets spat it in every direction.

Hollinsclough was taking a lot of water during that last hour. She sat very low, but with all that extra weight, she had become very stable. With *Clyde* on approach, we looked carefully at the life rafts — we were not saved yet. The safety of the rafts was not appealing. Their floors frozen with ice, we kept our fingers crossed.

Tracey stood tall, holding a mast steel. She was intently watching the pace of the rescue RIB battling towards us. "*Hollinsclough* is an English-built Oyster, and she is not going to let us down now." Sing it loud, we got on with the second verse of "All things bright and beautiful".

As the RIB closed, our last action was to drop the colours. Unclipping the mast stay, we rolled the ensign flag softly into itself. Wrapped tightly, Morgause attached it to the main recovery snatch bag. It was definitely going home.

As the yacht rolled and the deck dipped in the ocean, the Royal Navy manoeuvred the rescue RIB with perfection to land its bow softly wedged into our side rail.

Chief Engineer Lieutenant Rob Stavely stepped effortlessly from the RIB to our deck. He was aboard in a moment as the RIB withdrew for safety.

No survival suite, the lieutenant stood there in soft dark blue clothes and a life jacket. He turned to Tracey and said, "Lieutenant Rob Stavely, Royal Navy, how may we be of assistance?"

Wow!

Our whole bodies raced with excitement.

The situation was so desperate. We all stood perilously on the deck of a yacht about to sink, and Rob's formality assured us of success to survive this day.

Rob said, "Women and children first."

The RIB closed back, holding its bow as best it could against the *Hollinsclough*, the two rose and fell like the stairway in a funhouse. The floors twisted every one's feet in opposite directions.

Morgause judged the wave state well and leapt into the RIB. That's a girl who surfs a seven-foot board for you.

Morgause did not die today.

Caitland

First steps onto the warship

Caitland next. Rob held her by the arm as they both judged the next wave. Caitland landed sideways, but she was in.

Tracey took another wave set to be sure and leapt clean over the bow, where both girls took a hold of her.

This was no dinghy, it was a full diesel-powered monster RIB, but the sea state for its return was no better than its arrival.

Turning water tossed it about as it sped home to see the girls safe aboard *HMS Clyde*.

On board the warship, Lieutenant Helen took charge and delivered the team to the top deck suite of the executive officer, neighbour to the captain, and a new home safe from the waves.

There wasn't room on the RIB for all of us, and Rob had remained on the yacht with me.

We took time to get a clear view of the situation; I opened the deck windows

CHAPTER EIGHTEEN: *Hollinsclough Sinks*

This was no ordinary RIB boat.

to look down into the cabins below. It was a sight of carnage, floor panels floating about in a soup of saltwater and oil. Seats, cushions and bedding floated free to dance like croutons in the soup that was decidedly minestrone.

Amidst the swirling mess, the girls' schoolbook shelf was still above water in the centre cabin. Two handwritten work books stood out from the rest. There lay a whole year's maths work — revision summary, notes and all. Something so intense came over me. They were not for the ocean, and like the colours, they had to be saved. I returned down the stairway, the water was freezing, and the stink of battery rich in the last of the air space. I pulled the books free, turned back and handed them up to the Rob.

It was time to get off, and the RIB came back alongside. There was time to load every stuffer bag the girls had packed and even space for the grab canisters with the passports in. Life rafts were left inflated as a sign to the world that this was a stricken yacht.

Hollinsclough's own dinghy RIB floundered to her stern, steel rope rails broken with seas too high to successfully launch it. Morgause's work as a deckhand on the *Pharos*

Carl and Lieutenant Rob Stavely

A toy Morgause held dear to her heart, built with spares from around the world was her skateboard. The Royal Navy winched it to safety on its very own cord.

ship to twist the steel stay line was so strong and well twisted that it never broke. The trusty surfboard lay strapped solid to the stern davits. All ropes were in order and looped smartly to their winches as I took the last step off.

The RIB ride to the warship was motor against ocean as we battered our way against the wave tops. Once at the ship, a steel wire engaged a fast ascent winch as the crane sets tore the RIB from ocean into the air. We locked it into deck davits with a bang. White in the face, I could see Tracey and the girls safe aboard.

Naval traditions never to be forgotten, it was Caitland's idea.

I had rescued the finest bottle of champagne from the hold. All aboard decks to meet Lieutenant Commander Steve Moorhouse, skipper of the warship, I hugged him and handed over the champagne as a gift of honour to our rescue.

It was one o'clock, and we were family reunited in the XO's suite. It was a world apart from the environment we were accustomed to.

The chef had sorted meatballs and pasta, steaming dishes. Lieutenant Helen had more mugs of hot chocolate waiting, and there were tears in our eyes at the disbelief that we had been rescued from certain death.

There was a familiar face on the warship. British Antarctic Survey team doctor Susan Woodward was aboard — blue mafia, a fellow Girl Guide to the girls. She stood with us back on the bridge to watch the inevitable.

Hollinsclough lay very low in the water, her side windows submerged and the blue topside marker line under the waves. She held on steadfast and upright to the very end.

The decision was made to sink the life rafts. The confusion of faraway floating life rafts was uncomfortable, so Lieutenant Commander Steve Moorehouse despatched a gunner on the foredeck of the warship. He used his GPMG machinegun to sink the orange boats of safety — frozen floors and all. Tracer flash flew through the air with the smell of cordite. The wave state caused some bullets to ricochet into the air, deflecting off the banks of water that still twisted in the settling ocean.

CHAPTER EIGHTEEN: Hollinsclough Sinks

Goodbye to our home, she was a member of the family

HMS Clyde saluted with a long single blast of the warship's horn. Family together, we turned away as she went down.

There was 270 on the dial, all west for the Falklands.

One moment we were perilously close to the ice water of the Southern Ocean, the next we were plucked from the waves and all aboard the *HMS Clyde* with a meatball and pasta dinner. We had steamy plates and steamy clothes as the warmth around us de-chilled our bones. What's that smell? "Just might be us," said Caitland with a warm smile. Thirteen years old, and she had still never touched a glass of champagne.

It was hot showers all round with lashings of hot water — as much as we could use.

Sadly, we didn't have a toiletry to our name, as the aft heads was the first room flooded and never got packed.

Soap, conditioners and even a razor for me were rallied up amongst the crew.

Not only were we rescued, we had become part of a family. We were family of the Royal Navy Warship *HMS Clyde*, and what a family of friendship it was.

Sunday service was in the Officers' wardroom, and it was time for prayers. Lieutenant Helen played flute. The first hymn was "Eternal Father", and when we got to the end of the first verse with the line "those in peril of the sea", Tracey was in tears.

The whole family hugged. What an ordeal we had survived, and what a wonderful team of people had helped us in our survival.

A Southern Ocean Rescue is news, even bigger when it's an English warship rescuing an English yacht. Monday's *Sun* newspaper ran "Royal Navy rescue for Brit Boat that hits Titanic Iceberg".

There were no messages outbound, but support was still getting through, and it focused on the children. Morgause's best Repton school chum Ben and his mum who was in the Army, got a message in and words from school. That really touched little Morgause as she began to deal with the near-death questions of a pre-teenager escape. She held the maths book in her hand and said, "I will be back for end of term."

A message came in from Sue Bucket, FIC Falklands, our shipping agent chum who had so strongly supported us alongside at Stanley. "Will organise all logistics and accommodation, see you soon, God bless."

How our lives were in the hands of others.

Our travel of the world had been founded on people and friendship, and this was their finest hour.

It was clear to us that we owed our lives to three things:
~ The Royal Navy.
~ The strength of our English-built Oyster to last as long as she did after that ice impact.
~ The technology of communications.

And most of all to God. Bad luck rained about at the end of our adventures, but it was tremendous good fortune for *HMS Clyde* and her crew to have been down there in the Southern Ocean convergence zone of the South Atlantic Ocean.

It was surely a miracle to have been saved from such a place.

Epilogue

Journey from the Ice - The Way Home

Rescued from the Southern Ocean, we were on the south side of the world. It was a long way home, and in itself a story to tell, a story of friendship and support.

HMS Clyde, having rescued our family from the cold depths of the Southern Ocean, now low on fuel, steamed back to Mare Harbour on the Falklands. On the journey home, there was an iceberg collision drill to give us focus, but Lt Commander Steve Moorehouse and his crew made the girls feel like a part of their family. They baked cakes for the crew in the galley and practiced rope work on the deck.

Arriving in the Falklands, it seemed we were looked after by the whole island, fisheries ship *Pharos*, Chris and his crew awaited; it was a heartfelt tale of survival to share between sailors of these faraway oceans. Mike and his family took us in, organising our accommodation and giving great support to us in Stanley.

The Family on the London Eye in 2013

There was no quick exit, as all civilian flights were fully booked months ahead. With the help of the governor, we were found cancellation space on the military air bridge route to Ascension Island.

Stepping out into the sunshine, it was time to rest and recoup in the Hotel Obsidian, discover the land of Cable and Wireless and meet the astronauts. NASA kept a landing strip for the space shuttle here, and the BBC earth station relay sat on the highest hill. Again friendship was all around us, and there were even Girl Guides to say hello.

From Ascension, the planes had green paint, and we landed at Brize Norton with many soldiers returning from the war.

Our war was over, the girls would be back in Repton School within the week, and we hope you will join us for the next book: *Journey from the Ice, the Way Home from Certain Death*.

Acknowledgements

The sailors and friends of all the world who unreservedly took our ropes and tied us safe.

Girl Guides & schools around the World for the camaraderie with the girls, Father Hernan, the Jesuit priests of Chile and the Carmelite Nuns who prayed for us in our darkest hours.

Thank you to a team we couldn't have done without and those who prepared us for the ocean.

Our sailing began with powerboats, a foundation of training with the Royal Yacht Association sailing courses, Mark Goodwin tutored our sea survival at the Royal Marines SBS facility in Poole. Our time under power focused on Normandy, Brittany, Biscay and the Channel Islands. We learnt tides with the help of the Alderney islanders and sailed under the colours of the Royal Channel Island Yacht Club.

Rupert Knox Johnson sold us our yacht. Perfectly laid out for blue-water cruising, its strength proved a key factor in our sailing adventure. Ralph Catchpole was an inspiration on the final fit out, and ropes and standing rigging advice from Richard and Michael were invaluable. Well done the Barden team for power and charging.

Father Paul Baggot blessed the yacht with help from Ron and the marina team in St Katherine's London. Andrew Smythe at Cactus Marine fitted out the Iridium satellite communications that would be crucial, Ed Wildgoose at Mailasail organised the data transfer protocols. *Wandering Dragon* yacht helped us with our email & first Ugrib downloading. Bob and Di brought thoughts of communication code for dealing with pirates.

Derek Wood back in the UK was the fixer for all needs shore side from our teenage daughters' air travel to their boarding school pick-ups out of Repton, endless luggage balancing between spare parts and chocolate Easter Eggs. Well done Mrs Wherry for the half terms we were too far away for the girls.

John Kenyon MBE, who organised transit papers for Chile & Cape Horn.

Mike McLeod routed weather from the Falklands in our final hours.

Roy and Jen Dunning, sailing gurus for technical things and on watch all hours to the end.

Heavy-weather sailing introductions and discussions of power into the back of the mast were inspired by Jim Kilroy of Sydney Hobart *Kialoa* fame. We met Jim at the Punta Yacht Club in Uruguay, and his enviable experience was the foundation to our big-wind tactics.

We received some very special inspiration from crews around the world. Bob on *Penguin Two* taught the girls the art of lighting a Tilly lamp they would never forget. Ron and Dawn Moore from Jan Fiz focused on engineering, Thierry and Evelyne, Cool Daddy, submariner extraordinaire for underwater windows. The many marina chums who ferried shopping and stores across marina boards.

The crews and passengers of ships also played a key role of friendship as our travels became more distant—from small fishing boats gifting tuna to a cruise liner organising marzipan for our Christmas cake.

The Navy officers and crew of the world, especially Chile, Captain Pablo Muller and the teams at navy training school of Escuela de Grumetes, particularly Sgt Carlos Ojeda Guzman.

One ship of very special note is the *Pharos*, the Big Red Ship. Chris and his crew made us family in the most faraway waters in the world.

Annex 1 – A Tour of the Yacht

Our Home on the Water

Hollinsclough was an English built cutter rigged mast head sloop, with the international call sign VQRT8. We gave a tour of her cabins during the goodbye party at St Katherine's Haven, London.

Father Paul Baggot then blessed her before leaving the Haven.

Powered by a trusty blue Perkins diesel called Polly, two 2kw pumps to drive hydraulic pressure for the winch and furling sets of the sail wardrobe. There was a Westerbeke generator for 220volt power and a lot of locker space inside and out for proper live-aboard conditions. No gas, she had a diesel oven and hob. In the kitchen, an electric fridge, watermaker and washing machine resided.

We had VHF and SSB radio plus an Iridium satellite phone connected to data via the Mailasail network for world communication. Three EPIRBS were aboard for distress and two life rafts.

Focused on the best use of electricity, all lights were LEDs, even the navigation lights port and starboard. Solar panels helped top up the systems, and an aerogenerator maximised charging when the wind was strong. A giant bank of heavy six-volt batteries stored the power in two systems, one 24 volt and one 12 volt.

We had a whole host of maps and a fabulous cockpit compass, but primary navigation was by Raymarine chart plotters. There was a 4kw radar on the mast to see in the dark, many spare navigation systems, and a handheld satnav was kept inside the spare microwave oven—it was protected even from lightning strike.

CERTAIN DEATH IN THE ICE

Hollinsclough was steered from the centre cockpit, which had enclosed covers for the cold weather. Her main bedroom was to the rear stern, with the girls' bedrooms forward. There were two toilets and showers that would both help pump water when our need was greatest.